T0300266

Routledge Revivals

The Circumpolar North

First Published in 1978 *The Circumpolar North* is designed for anyone with a more than superficial interest in the northern regions of our planet, geographical, economic, social, or political. The primary importance of North today is as a source of raw materials, as a world crossroads, and as a touchstone of the way nations behave towards their minority groups. Strategic considerations have led to the expenditure of vast sums of money; but world population expansion has not yet affected the northlands and their preservation in a natural state is still a feasible objective.

The authors are experts in their own areas and have provided regional chapters on each of the land and ocean areas. The book compares the different approaches of the countries involved and deals also, in the context of the northern seas, with another political dimension – the relations between nations and their success in achieving international management of resources. This is an interesting read for scholars of geography, international relations and international economics.

The Circumpolar North

A Political and Economic Geography of the Arctic and Sub-Arctic

Terence Armstrong, George Rogers, and Graham Rowley

Routledge
Taylor & Francis Group

First published in 1978
by Methuen & Co Ltd.

This edition first published in 2023 by Routledge
4 Park Square, Milton Park, Abingdon, Oxon, OX14 4RN

and by Routledge
605 Third Avenue, New York, NY 10017

Routledge is an imprint of the Taylor & Francis Group, an informa business

© 1978 Terence Armstrong, George Rogers, Graham Rowley

Publisher's Note
The publisher has gone to great lengths to ensure the quality of this reprint but points
out that some imperfections in the original copies may be apparent.

Disclaimer
The publisher has made every effort to trace copyright holders and welcomes
correspondence from those they have been unable to contact.

A Library of Congress record exists under ISBN:

ISBN: 978-1-032-45326-2 (hbk)
ISBN: 978-1-003-37908-9 (ebk)
ISBN: 978-1-032-45371-2 (pbk)

Book DOI 10.4324/9781003379089

TERENCE ARMSTRONG
GEORGE ROGERS   GRAHAM ROWLEY

The Circumpolar North

A POLITICAL AND ECONOMIC
GEOGRAPHY OF THE ARCTIC
AND SUB-ARCTIC

METHUEN & CO LTD   LONDON

First published in 1978 by
Methuen & Co. Ltd
11 New Fetter Lane   London EC4P 4EE
©*1978 Terence Armstrong, George Rogers, Graham*
Rowley
Printed in Great Britain at the
University Press, Cambridge

ISBN *(hardback)* 0 416 16930 9
ISBN *(paperback)* 0 416 85430 3

Contents

viii

Maps

Tables

xi

Glossary of terms

Active layer, see **Permafrost**

Boreal forest, see **Taiga**

Fast ice. **Sea ice** which remains fast along the coast, where it is attached to the shore or to shoals.

Iceberg. A large mass of floating or stranded ice of greatly varying shape, more than 5 m above sea level, which has broken away from a glacier.

Ice cap. A dome-shaped glacier usually covering a highland area. See **ice sheet**.

Ice island. A form of tabular **iceberg** found in the Arctic Ocean, with a thickness of 30 to 60 m and from a few thousand sq m to 700 sq km in area.

Ice sheet. A mass of ice and snow of considerable thickness and large area, resting on rock. Properly speaking, there are at present only two in the world: the Greenland ice sheet and the Antarctic ice sheet. Smaller features of the same kind are called **ice caps.**

Ice fog. A suspension of numerous minute ice crystals in the air, reducing visibility at the earth's surface. It may be produced by a continuing source of water vapour in calm air at temperatures of -35 °C or lower. The source may be industrial cooling waters or combustion products, car exhaust, or just human or animal exhalation. Strong inversion will tend to keep the fog in place.

Land ice. Ice found on land. See **Ice cap** and **Ice sheet**.

Muskeg. Peat bog, chiefly in the boreal forest. The term originated in Canada, and is not used outside North America. In the USSR the equivalent term is sphagnum bog. Muskeg is almost impassable to men or vehicles except when frozen.

Pack ice. Any area of sea ice, river ice or lake ice which is not attached to the shore (when it is known at sea as **fast ice**). Pack ice at sea is continuously moving, and for that reason is sometimes called drift ice. Level pack ice seldom exceeds 3m in thickness, but when the ice has been broken by pressure into ridges and hummocks it may reach thicknesses of 20m or more.

Permafrost. Soil or rock of which the temperature remains below 0 °C continuously for a year or more. Frozen ground almost always contains some ice, but this is not an essential constituent; in fact unfrozen water may be found in it, if impurities or pressure prevent it freezing. Permafrost may persist for only one year or for many thousands of years, and it may be a few centimetres or many hundreds of metres thick. The surface layer, which freezes and thaws seasonally, is called the active layer, and this too varies in thickness up to about one metre with climatic and terrain factors. The term permafrost, which was coined only in 1945, stood for permanently frozen ground. But it is better to substitute perennially for permanently, because the phenomenon can thaw and disappear. Many land forms and processes owe their origin to permafrost: patterned ground, pingos, frost boils, icings, and thermokarst topography in general.

Sea ice. Any form of ice found at sea which has originated from the freezing of salt water: the frozen surface of the sea. The term thus excludes icebergs, which break off from glaciers where these enter the sea, and are therefore of land ice origin.

Taiga (also spelt tayga). The vast, predominantly coniferous, forest lying immediately south of the **tundra** in Eurasia; effectively the same as the boreal forest, a term more frequently used in North America.

Tree line. The northern limit of trees; usually a rather broad zone. Different species have different northern limits, but the treeline as generally understood is the point beyond which only dwarf species grow. The term can also be used of altitude in mountain regions, where the line is often more clearly defined.

Tundra. Treeless plain, usually underlain by permafrost, lying north of the tree line. It is found in all Arctic lands, generally near the coast. Tundra has its own vegetation.

Wind-chill. The term was introduced by the Antarctic explorer Paul Siple in 1939 to signify the cooling power of wind and temperature combination on human skin which is dry and in shade. A formula was produced and later improved, and this is now used to indicate the degree of discomfort due to these factors at particular times and places.

Abbreviations

Abbreviations, apart from such standard forms as m for metre, have been in general avoided. But a certain number have occasionally been used, and these are listed below.

AO	Avtonomnaya oblast' (Autonomous Province) (USSR)
ASSR	Avtonomnaya sovetskaya sotsialisticheskaya respublika (Autonomous Soviet Socialist Republic) (USSR)
BAM	Baykal-Amur Magistral' (Baykal-Amur main line) (USSR)
BMEWS	Ballistic missile early warning system
Dal'stroy	Glavnoye upravleniye stroitel'stva dal'nego severa (Chief Administration of Construction in the Far North) (USSR)
DEW line	Distant early warning line
dwt	deadweight tons, a measure of the cargo-carrying capacity of a ship in tons weight
EEC	European Economic Community

Glavsevmorput'	Glavnoye upravleniye severnogo morskogo puti (Chief Administration of the Northern Sea Route) (USSR)
Gosplan	Gorsudarstvennyy planovyy komitet (State Planning Committee) (USSR)
GTO	Grønlands Tekniske Organisation
ICES	International Council for the Exploration of the Sea
ICNAF	International Commission for the Northwest Atlantic Fisheries
KGH	Den Kongelige Grønlandske Handel (the Royal Greenland Trading Company)
KPSS	Kommunisticheskaya partiya sovetskogo soyuza (Communist Party of the Soviet Union)
LNG	liquefied natural gas
NATO	North Atlantic Treaty Organization
NEAFC	North East Atlantic Fisheries Commission
NKVD	Narodnyy komissariat vnutrennikh del (Peoples Commissariat for Internal Affairs) (USSR)
NO	Natsional'nyy okrug (National District) (USSR)
RGTC	Royal Greenland Trading Company (see also KGH)
shp	shaft horsepower

Preface

This book is designed for anyone with a more than superficial interest in the northern regions of our planet. The authors have had particularly in mind its use as a textbook for courses at university or pre-university classes at school. The justification for publication now must be the rapid pace of change in the world, and therefore in its northern regions also. A great deal has happened in recent years: major new developments in the extractive industries, trans-Arctic flying on a routine basis, commissioning of nuclear icebreakers, statehood for Alaska, awakening of national consciousness in many northern peoples – to name only some of the novelties. Knowledge of the natural environment has also increased greatly, partly through the stimulus provided by the International Geophysical Year and other major international programs. The growth of knowledge has been so large that we decided to limit the cover of this work to political and economic affairs, with some environmental background, rather than to attempt equal treatment for all aspects.

The book has regional chapters on each of the land areas and on the oceans, and a concluding chapter viewing particular aspects or activities on a circumpolar basis. While the subject matter of the regional chapters tends to follow the same general plan, we have avoided any squeezing of the material into too rigid a pattern. The regions all have their special features, requiring special treatment. Thus the Soviet chapter is less fact-filled than some of the others, because many facts about the Soviet north are not published; the Alaskan chapter is fuller and larger than might be expected on grounds of area and population, because it so happens that data are abundant and offer the chance of detailing 'the evolution of a northern polity', as the chapter sub-heading puts it. Northern Scandinavia and Iceland are deliberately given less space, for their problems are both well covered in other publications and are also somewhat atypical of those in the other land areas. While a book of this sort must bear the marks of joint authorship, it is also the product of close collaboration, and it is fair to say that each author has made contributions to every chapter.

Acknowledgments are due to many friends who have helped in many different ways. Special mention must be made of Professor Trevor Lloyd, to whose initiative the book owes its inception. All the authors would like to express thanks to the Scott Polar Research Institute, Cambridge, for its encouragement and support, and for the use of its excellent library.

Place-names follow the joint usage of the British Permanent Committee on Geographical Names and the American Board of Geographic Names, as does the system of transliteration from Russian. Distances and weights are expressed in the metric system, with non-metric equivalents where appropriate. The metric ton is written tonne. Ton alone should be taken to mean long ton, which is almost exactly equivalent to a tonne and so needs no metric equivalent. Short tons are 10 per cent less in weight than tonnes or long tons, so the metric equivalent has been indicated where helpful. Billion and trillion mean 10^9 and 10^{12}. It is not appropriate in a book of this kind to identify sources for all statements of fact; but it is done in the case of tables, and the list of further reading includes major sources.

TERENCE ARMSTRONG
GEORGE ROGERS
GRAHAM ROWLEY

Cambridge, Juneau, Ottawa
June 1977

CHAPTER ONE

Introduction

The area and its importance

The circumpolar north is a convenient abbreviation for the Arctic and sub-Arctic. These can be defined in many ways, varying with the subject of investigation. If one is thinking of the Arctic as the place of the midnight sun, then the Arctic Circle (lat. 66° 33′N), which delimits the area within which the sun does not rise at the winter solstice nor set at the summer solstice, is a good boundary. If one is thinking of it as the treeless land, the boundary is the northern limit of trees, a very different line on the map. There is no lack of definitions. The difficulty arises when one tries to find a definition to suit all subjects. It is no good taking an average between several widely varying lines. It would be possible to take the lowest common denominator, and draw the southern boundary in such a way that every Arctic and sub-Arctic phenomenon were included. But this would enclose a very large area indeed: there is true tundra in Scotland, and the sea freezes at Odessa. So the most desirable solution is a flexible one: to think of the Arctic and sub-Arctic as a group of concepts and attributes, concerned with climate, vegetation, fauna, presence of ice and snow, sparseness of human habitation, remoteness from industrial centres, and many other factors, and not having precise boundaries.

Nevertheless, it is still necessary to have in mind some general limit, for the guidance of the reader. For the purposes of this book, therefore, the Arctic and sub-Arctic, taken together, are deemed to include: in North America, the northern parts of the Canadian provinces, including all

Labrador, and the whole of the Northwest Territories, the Yukon Territory, and the State of Alaska; in Eurasia, the territory north of about latitude 65° N in Scandinavia, with the boundary dropping to 63° in European USSR and 57° in Asia, and including all Kamchatka; the whole of Greenland, Svalbard, and Iceland; and at sea, the Arctic Ocean, with the Barents Sea, the Norwegian Sea (including the Greenland Sea), Denmark Strait and the Labrador Sea on the Atlantic side, and the Bering Sea, the Sea of Okhotsk and the Gulf of Alaska on the Pacific side (Map 1).

The boundary between the Arctic and the sub-Arctic could be dealt with in the same way, but since the two regions are adjacent, the southern boundary of the one being the northern boundary of the other, and since this book is concerned with both of them, there is little to be gained by doing this. The distinction between them is in most cases one of degree. But in any consideration of economic matters, it is helpful to group the two together, because the sub-Arctic is where the present frontier of settlement and development is found in many parts of the world, and consideration of what happens there, and in what circumstances the frontier may be advanced northwards, is the principal subject of the book.

These northern regions have an area of about 41 million sq km, and constitute about 8 per cent of the surface of the planet. The lands within them are about 15 per cent of the land area of the world, and the seas about 5 per cent of world ocean. These are significant proportions. Yet the human population is under 9 million, or about 0.3 per cent of world population. The eight countries that claim sovereignty over the land are among the most advanced in the world from a technological point of view: in descending size of northern territory, the USSR, Canada, Denmark, the USA, Norway, Sweden, Finland, and Iceland. The 'sovereignty of the seas' is still emerging in patterns of multi-national treaties and arrangements, and concepts of international law are becoming conventions adhered to by all.

There are a number of ways in which interest in these regions is becoming apparent. The most important at this moment is the search for, and exploitation of, raw materials. One might reasonably expect that 8 per cent of the surface of the planet would contain 8 per cent of the minerals in the earth's crust. Geological exploration so far provides no evidence that the northlands have less than their expectable share. The Russians have gone farther than anyone else in this direction. Powerfully motivated by the desire for national self-sufficiency, they have sought and found much, including very important deposits of gold, nickel, tin, diamonds, oil and gas. The search and the exploitation will certainly continue. The resources in question, it should be noted, are non-renewable. It is inherently less likely that there will be exploitation on the same scale

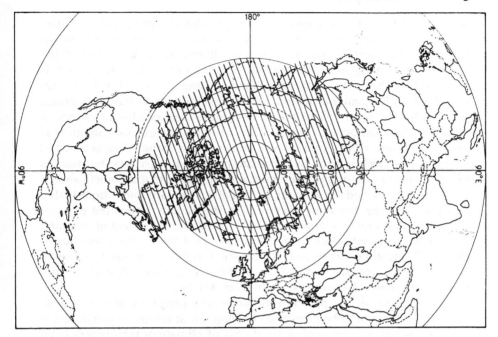

Map 1. Area covered by this book

of renewable resources, for reasons of climate and soil; but this does not mean that renewable resources cannot be found and used at all. Fish, fur, and timber, to list only three, are already harvested in the north. All maritime nations have a present or potential share in the harvest of marine resources.

The importance of the north as potential living space is not yet significant. Although world population is increasing at a frightening rate, it is still true to say, in the 1970s, that every human being could stand (but not sit!) on the Isle of Wight, or even Martha's Vineyard. Long before these world-wide trends were manifest, each of the nations acquiring northern territories had included among its stated objectives the occupation and settlement of empty lands and the political and social, as well as economic, advancement of the resident population. At times investment of money and organization of programs have backed up the words of these statements, but a range of problems has limited their fulfilment.

The problem of northern settlement is not simply a matter of where people can physically be, but of how to provide the essentials of life for them. Local resources can support only the native population, with no large surplus to feed the recent immigrants. The climate is frequently harsh and northern location means abnormally long periods of continuous darkness or light. Remoteness from the populated centres and amenities of the south represents a real psychological barrier. Movement of population into the north is likely to happen, therefore, only as part of the process of making available the north's resources. The likelihood is increased by the fact that the northlands belong to the rich nations, not the 'third world', and they do not have to develop broadly based economic systems for the north. Since they can operate their northern territory as a part of their national economy, their attitude towards development is likely to be one which stresses use of resources rather than settlement potentialities. This is perhaps fortunate because it increases the chances of taking into account another, more recent, view, that the north should be as little disturbed by humans as possible. The argument here is that it is precisely the mounting pressures in the already populated parts of the world that create a critical scarcity of havens of essential solitude, natural beauty, and absence of pollutants. That these characteristics of the north must be preserved is already forcibly expressed by many Americans with regard to the State of Alaska, and we will undoubtedly hear plenty more of it, and not only over Alaska.

Manifestation of the problem of centre and periphery and the whole range of settlement difficulties and challenges that relates to geographical location, whether in respect of harshness of climate or remoteness from population centres, will also be the special concern of this book. Here Canada merits particular attention, as a northern nation in a more important sense than the others: it is the second largest country in the world, and three-quarters of it lie within our area of interest as just defined. Canada, more than the United States or even the Soviet Union, has sought its national identity by looking north and is the most likely to 'move north'.

A Pole-centred map of the northern hemisphere demonstrates at once that the Arctic Ocean is a mediterranean sea, and, if suitable modes of transport are available, the shortest routes between many of the world's major population centres (which happen to be almost all in the northern hemisphere) pass across it or close to it. The old dream of an Arctic short-cut between east and west is beginning to be realized. Aircraft of many countries fly regularly across the Arctic Ocean, and ships may soon be able to follow a similar route. The north has growing importance as a transport nexus.

There remains another way in which the northlands can be, and have been, put to use by man. This is as strategic space, and no nation with territory there has been able to ignore this aspect. After the Second World War, the two major protagonists in any future conflict emerged as the USA and the USSR; and the shortest routes between them, either absolutely or as measured between major industrial regions on each side, lie across the northlands. At the same time weapons became available which could cover the distances involved, at first long-range bombers, then ballistic missiles. This led to construction, at immense expense, of airfields and chains of radar stations. Quite apart from the directly military significance of all this, the program involved so much building of infrastructure, training of personnel, and undertaking of pure and applied research that many non-military activities were affected by them. In recent years, further increases in the range and capabilities of weapons systems have in some respects reduced the strategic importance of the northlands; it is no longer so necessary to concentrate on the shortest distances between the two countries, and in any case China, with no Arctic frontier, may be joining the other two as a major protagonist. But certain aspects remain important: empty territory can pose strategic problems and an Arctic location has strategic advantages in the air age.

These reasons for taking an interest in the north – extraction of raw materials, provision of living space, expansion of transport systems, and use as strategic space – relate to the situation viewed from the outside. But there are 9 million persons already living in the north, and about a million of these are indigenous peoples, such as the Eskimos and Lapps. The average density of population is very low – 0.4 persons per sq km – but in absolute terms 9 million is no negligible figure. These persons are nationals of the eight countries concerned, in each of which (except Iceland) they constitute a minority, often a very small one. The question of relations between the paramount power and its northern subjects, particularly the natives, is one that has acquired much importance in the last few years. In North America and Greenland, it is now the most important question of all, having gained priority in Canada over the extraction of raw materials, and giving rise to major advance in the political status of the people of Alaska and Greenland. This may seem to be a purely internal aspect of northern affairs, concerning the great world outside scarcely, if at all; but this is not so. The relationship illustrates certain points of principle in such areas as justice, education, and welfare which could be of general application. The eyes of the world are upon the northern peoples and the way in which they are treated.

Thus the primary importance of the north is as a source of raw materials, as a world crossroads, and as a touchstone of the way nations behave towards their minority groups. Strategic considerations have led to the expenditure of vast sums of money, from which certain non-military advantages were nevertheless gained, but the importance of this is now diminishing. The time has not yet come when the northlands need to be used for living space in an overcrowded world, so their preservation in a natural or near-natural state is still a feasible objective.

Beyond being a study of the contemporary northern lands and seas in transition, this book also aims to be a case study in regional development and political and economic change which may illuminate these phenomena in more complex settings than the north. As already noted, the sovereignty over the lands and territorial waters is divided among eight nations. Together they represent a range of differing political and social institutions and economic systems and objectives. A comparison of how these essentially southern phenomena have been transplanted, and function or malfunction in the shared physical environment of the northlands, can contribute to an understanding of them in their national setting. These national divisions provide a catalogue of political entities in differing degrees of evolution: the independent Republic of Iceland, the nationally integrated State of Alaska, Greenland on the threshold of home rule, the territories of Canada (Yukon and Northwest Territories), and the northern administrative divisions of the USSR, the Canadian provinces, and Scandinavia. Comparative study of these units in their different stages of development can provide insight into the process of political evolution from contemporary sources as a supplement to the more usual historical approach. The different political environments also provide a means of studying the non-economic aspects of economic change and development. Turning to the seas of the north, this study will be dealing with another dimension and level of politics – the relations between nations (not limited only to the north), conflicts and co-operation, and the extent of ways and means to achieving international management of resources.

The geographical background
Land and sea

The northern part of our planet is an ocean surrounded almost completely by land. The land is the northern edge of two major land masses – Eurasia and North America – with many off-lying islands, especially on the American side. The only substantial break in the ring of land is between Greenland and Scandinavia. Here, on the Norway side of the opening, the North Atlantic drift brings the main inflow of water into the Arctic Ocean, and on the Greenland side, the East Greenland current constitutes the main outflow. There are other breaks: Bering Strait, the shallow 90-km gap between Asia and America, and the generally rather narrower straits between the islands of the Canadian archipelago. The net flow is into the Arctic Ocean through the Bering Strait, and out of it through the archipelago.

The American northlands have as their principal feature the great Canadian shield, a rock mass of Precambrian age which occupies a wide stretch of land centred on Hudson Bay, including the whole of the Labrador peninsula and the southeastern islands of the archipelago. Greenland is partly an extension of the Canadian shield, but the northwest is related to the northern islands of the archipelago, and the east coast has Tertiary basalt plateaus reminiscent of Iceland. Greenland is therefore in the nature of a bridge between adjacent regions.

The northlands of Eurasia may be subdivided in an analogous way. There are two, considerably smaller, shield areas: the Baltic shield, and the Anabar shield in north-central Siberia. Midway between them are the Ural mountains, and on either side of that range there are extensive lowlands. The largest of these, and an important feature in its own right, is the west Siberian lowland, lying between the Ob' and Yenisey rivers. The horizontally bedded sediments are of very great thickness, and have a surface area of 2 million sq km. Drainage is bad, and much of the region is swampy. South-central and eastern Siberia, and also the Pacific coast, are areas of folded mountains, of which the highest (up to 4750 m) are in the peninsula of Kamchatka. That area is highly seismic and volcanic, a characteristic it shares with other parts of the North Pacific rim, notably southern Alaska.

That is a very broad description of the situation as it exists today, and has existed in historical times. But it has not always been so. In the geological past, the continents have not remained where they are now, either in relation to each other or to the North Pole. They have moved around on the surface of the planet, like rafts in a viscous sea. One hundred million years ago southeast Alaska was at the Pole. Earlier than that northeast Siberia was there, and earlier still – 500 million years ago – none of the land masses at present surrounding the Pole were within 5,000 km of it. So all land areas at present in the Arctic or sub-Arctic have at some time in the past been in low latitudes, and may have had a climate very different from their present one.

The Arctic regions were known to man long before recorded history gives us any written record of the facts. Archaeological studies show the existence of ancient sites at many points on or near the shores of the Arctic Ocean. In North America the Eskimos spread from west to east across the top of the continent, reaching around Greenland, to produce a cultural uniformity along the whole Arctic coast. In Siberia, where the main river systems run northward, the different ethnic groups look south for their origins, and there is a wide diversity in language. The diversity in culture is not so marked because the similarity in conditions and resources has often resulted in similar techniques and similar solutions to problems. The different groups have lived side by side for so long that there has been a considerable amount of borrowing of cultural traits, with complete absorption of some of the smaller groups by the larger.

Relating the distribution of land and sea to the present political situation, we may note that Greenland and Alaska are separated physically from the rest of the nations of which they form part. This has resulted in each having a large measure of internal political, economic, and administrative homogeneity. In contrast, the Soviet and Canadian northlands have been penetrated by southern influences from many points and in many different ways. Greenland and Alaska have a well-defined boundary; the Soviet and Canadian northlands have had no clear border to protect them or to be defended.

Snow and ice on land

The popular impression of the Arctic and sub-Arctic as lands of ice and snow is a considerable exaggeration. Snowfall can be heavy in some areas, notably on the Pacific and Atlantic coasts of the sub-Arctic, but in general it is light. Along the Arctic Ocean coasts the winter accumulation is mostly under 25 cm, and for the Arctic as a whole it rarely exceeds 75 cm. In the summer, snow disappears almost everywhere. It may lie for up to nine months in the year, but many places in central Canada or Siberia have much more snow although it lies for six months or less. Snowfall in the north appears to be more frequent than it is because the snow is often picked up and carried by the wind. Lying snow is rarely a serious obstacle to human activity in the north, but blowing snow can be.

Land ice, in the form of ice sheets, ice caps, and glaciers, is a more formidable feature. It covers about 2.3 million sq km in northern regions (see Map 2). Over four-fifths of this area of ice is in Greenland, where it is called an ice sheet, a term otherwise used only for the Antarctic ice sheet. Outside Greenland, the land ice is found mostly on the northern and eastern islands of the Canadian archipelago, and on some islands north of Eurasia – Svalbard, Zemlya Frantsa-Iosifa, Novaya Zemlya, and Severnaya Zemlya. On these islands there are quite extensive ice caps, many valley glaciers, and in one or two places ice shelves (land ice which is afloat, yet attached to the land). A number of glaciers come down to the coast, and discharge icebergs into the sea.

Almost all land ice is moving, and the rate of advance can be a sensitive index of climatic change. There have been too few observations in the Arctic and sub-Arctic, and they cover too short a time, for any reliable comprehensive statement to be made about variations in extent. It is clear, however, that there were advances in many areas in the eighteenth and early nineteenth centuries, while today many Arctic glaciers are retreating. The full picture is evidently complex, for it is possible to find, for instance in Alaska, an advancing and a retreating glacier within a few miles of each other. There can be direct practical implications in glacier movement. An advance may block road or rail communication, a retreat may cause boundary problems, as on the US–Canadian border in Glacier Bay. So-called glacier surges, or catastrophic advances, can be still more

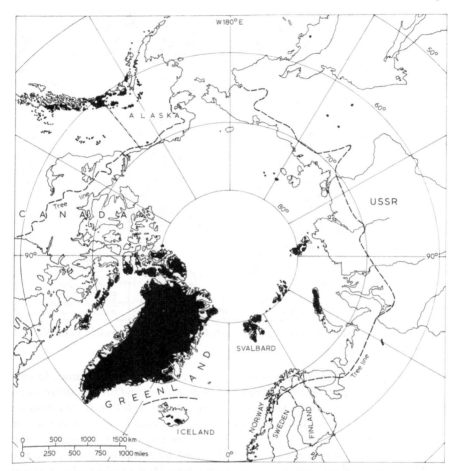

Map 2. Land ice in the circumpolar north. From R. P. Sharp in *Polar Record*, vol. 9, no. 58 (1958), pp. 34–5

damaging. But the chief limitation imposed by the presence of land ice is, of course, the difficulty, amounting effectively to impossibility, of getting at the land underneath. Thus almost the whole of Greenland, which presumably has mineral wealth commensurate with the size of the island, is denied to geological exploration and prospecting.

A third manifestation of a cold climate is the presence of permafrost, or perennially frozen ground. Permafrost lies beneath 20 per cent of the land area of the world, including about half the total territory of Canada and the USSR. The soil on and immediately below the surface thaws in the summer, but below this so-called 'active layer', which is up to 2 m thick, there is a perennially frozen layer which may be more than 1,000 m

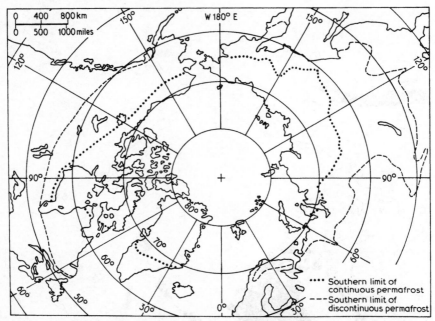

Map 3. Permafrost in the circumpolar north. From R. J. E. Brown in *Polar Record*, vol. 13, no. 87 (1967), p. 745

thick. In general the thickness diminishes as the latitude becomes lower. In the northern part of the permafrost zone the permafrost is continuous, but farther south it becomes discontinuous, that is, it is found only in some localities, such as north-facing slopes, or as small bodies in particular situations, and the ground between these is unfrozen. The southern limit of the discontinuous zone is therefore hard to determine, but certainly reaches south of latitude 50° N in places (see Map 3). Permafrost is not generally present under water, but it has been found in the shallow waters off the coast of Siberia and in Mackenzie Bay.

The effects of permafrost are often visible on the surface. Patterned ground, pingos or frost boils, and thermokarst topography are examples. Since the permafrost layer restricts drainage, moisture from precipitation is retained near the surface. This results in solifluction, mud flows, and landslides on slopes, and may provide conditions for more abundant vegetation.

This effect on drainage also has many far-reaching consequences when structures are built on or in the permafrost. Almost any interference with the natural state changes the thermal conditions and is liable to alter the position of the permafrost table, and therefore of the load-bearing capacity of the ground. Frost heave, caused by movement of ground water, or differential melting brought about by the conduction of heat through foundations into the permafrost, have caused distortion of buildings and, in some cases, their total destruction. Two very important considerations when selecting the site for a building are therefore these: to find ground which contains as little moisture as possible – bedrock best of all – and to pay very particular attention to surface drainage. Interference with the permafrost should be kept to a minimum, unless the permafrost at the site is thin and in the discontinuous zone, when a better policy may be to melt it out altogether before starting construction. Much work has been done in both North America and the USSR on the natural history of permafrost as well as on its engineering aspects, and its behaviour in most situations is now quite well known and predictable. The measures necessary to counteract the harmful effects give rise, of course, to extra expense, and these are one of the many factors which make any undertaking in the north more expensive than it would be in temperate latitudes.

Floating ice

The presence of floating ice is one of the main physical features of the region, affecting profoundly many aspects of its natural history, and incidentally creating a major difficulty for ships. Two main kinds of floating ice may be distinguished: first, the frozen surface of the water body, called sea ice, lake ice or river ice according to where it was formed; and second, land ice which has reached water and floated off in the form of icebergs or ice islands. Well over 95 per cent of the floating ice found in the north is of the first kind, but the second is important because it constitutes a special danger to shipping.

Sea ice (see Map 4) may cover an area of 15 million sq km at its greatest seasonal extension in early spring, but only just over half that area at the peak of the melting season in late summer (these figures leave out of account varying concentrations of ice, i.e. the water areas enclosed within the ice). The main body of the Arctic Ocean is always ice-covered, but the ice gets somewhat looser in summer, with more cracks and leads. This very large variation between maximum and minimum extent is paralleled by a less large, but still very considerable, variation between the maxima and the minima of different years. Variability, which is controlled by a complex interaction of meteorological and oceanographic factors, is therefore a significant characteristic.

Sea ice may be of varying thickness and strength. The initial stages of growth may be a coagulation of ice crystals to form a soupy layer on the surface, called grease ice. This may turn into spongy white lumps (shuga), or a thin elastic crust (nilas), or a brittle, shiny crust (ice rind). Each of these will consolidate with growth, and during the first winter of

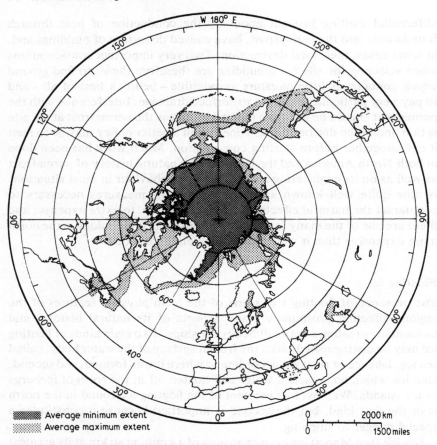

W 180° E

Average minimum extent
Average maximum extent

0   2000 km
0   1500 miles

Map 4. Sea ice in the northern hemisphere. From *Map of the Arctic region*, 1:5,000,000, American Geographical Society, 1975

existence may attain a thickness of up to 2 m (first-year ice). Such ice will offer some resistance to ships, and will also carry a considerable load. This is the predominant ice type in sea areas which wholly clear each summer, and there is a fair proportion of such ice in the Arctic Ocean also. The remaining sea ice is older: second-year ice, or after two years, old ice. This ice is thicker, and also disproportionately stronger, because most of the brine has by now been leeched out. An undisturbed level floe may approach a thickness of 3 m, but in practice floes of this age have almost always been broken by pressure into ridges and hummocks, and the refrozen and consolidated fragments may reach thicknesses of 20 m or more. This is the dominant ice in the central part of the Arctic Ocean and the northern part of the Canadian archipelago, but some old floes drift southwards and may be found in the peripheral seas also.

Almost the whole of this mass of sea ice is in continuous movement, responding to the action of winds, currents, and Coriolis force, and is known as pack ice. There are two principal circulations which are constant in general direction, though varying in speed. The first is from points north of Bering Strait and north of the Siberian coast across the centre of the Arctic Ocean towards the Greenland Sea, where the ice is borne southwards to melt in the North Atlantic. A given floe travels at 2–3 cm/sec, and might take two to four years to complete the traverse. The second is a closed clockwise circulation in the waters north of Alaska and of the Canadian archipelago, known as the Pacific gyre. A full circuit takes about four years. Because of the slow rate of drift, and the fact that much of the ice cannot escape from the closed system, the thickest ice in the Arctic Ocean is to be expected here. However, it is unlikely that an average thickness of much more than 7 m will be found (few detailed measurements have yet been made), because sea ice ceases to increase in thickness when it has grown to the point at which it effectively insulates the water below from the cold air above.

The only substantial part of the sea ice which is not in movement is the fast ice, which is held in place by the shore and remains relatively immobile. Fast ice commonly fills sheltered bays and straits, but may extend as far as 500 km from the shore of the mainland off parts of northern Siberia, where the continental shelf is particularly wide. Almost all the fast ice is first-year ice, and clears away completely in the summer.

Lake ice and river ice are simpler phenomena than sea ice, in that they, with rare exceptions, do not survive into a second winter. When they melt, or are borne downstream in the spring, water remains clear of ice until autumn. There is also somewhat less variability in the dates of break-up and freeze-up.

The second major kind of floating ice is ice of land origin. Icebergs in the north may be of considerable size: a large town square, or a city block, might be the average. They can be over 70 m high, and the draught under water is generally about five times the height, depending on a number of factors. As they melt, they break up into smaller pieces known as bergy bits (the size of a house) or growlers (the size of a car). The greatest number of icebergs in the north is found in Baffin Bay. They calve from the glaciers of west Greenland, and are carried across Baffin Bay and down into the North Atlantic, where most finally melt off Labrador and Newfoundland. It is here that they endanger shipping lanes. Icebergs occur in other regions where glaciers reach the sea, but are not so numerous and usually remain close to their source. One special type should be mentioned, though it is not very common: the flat-topped tabular iceberg, known in the north as an ice island. These originate from ice shelves in northern Ellesmere Island, and are found in the Arctic Ocean and the waters of the Canadian archipelago. They are generally about 50 m thick, and may be as much as 700 sq km in area.

Floating ice is of course a major obstacle to ships. A vessel must be specially strengthened to work in ice at all, and must be specially designed if she is to take a leading part. In the present state of icebreaker development, first-year ice is about the limit of what can be broken without considerable difficulty; but it does not follow that all first-year ice, wherever and whenever encountered, presents no problem. Older ice is progressively more difficult, and icebergs are impossible (but can normally be avoided). No ship has yet been built which can go anywhere at any time in ice-filled waters. However, it does now seem possible, though not certain, that such a ship could be built if enough money were available. So the whole question of navigation in sea ice is at an interesting threshold. Submarines, of course, largely avoid these difficulties, though icebergs present a hazard, as do the shallow waters of the continental shelf. Naval submarines have amply demonstrated that most parts of the Arctic Ocean can be reached. Freighter submarines have been designed, but here again, the building of such ships awaits economic justification.

Climate

There is wide variation between the climate of places at the same latitude. At latitude 60° N on the European side of the Atlantic lie the Shetland Islands and the rainy southwest coast of Norway, while on the North American side there is the almost wholly ice-covered tip of Greenland. The Arctic Circle is quite misleading as an indicator of climate. The main warming influence coming into the Arctic is the North Atlantic drift, a major ocean current which enters the Arctic Ocean through the Norwegian Sea, and spreads its waters to Svalbard, the Barents Sea (where it causes Murmansk, the USSR's most northerly port, to be also the only permanently ice-free major port in the country), and, below the surface, to very large areas of the Arctic Ocean. Conversely, important cooling influences emanating from the Arctic are the East Greenland current, with its burden of sea ice, the Labrador current, which bears icebergs as well as sea ice into the North Atlantic and perhaps most important of all, the outbreaks of Arctic air, particularly in Siberia and North America in spring and summer.

The North Pole is by no means the coldest place in the north, since its altitude is low and the sea is a moderating influence. The mean annual temperature there has been estimated to be −23 °C (no recording station

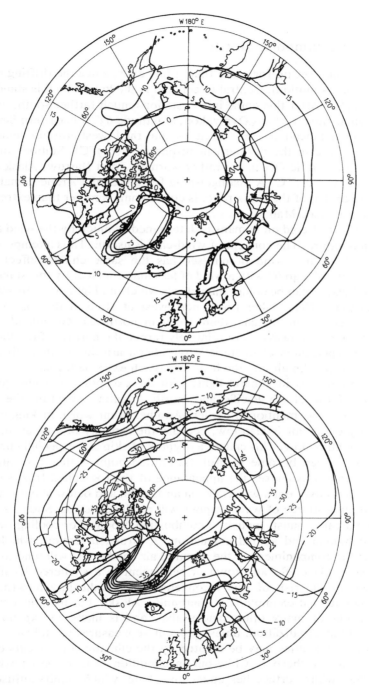

Map 5. Mean surface air temperatures in the circumpolar north, August and February. From G. A. McKay and others in *Proceedings of the conference on productivity and conservation in northern circumpolar lands, Edmonton, October 1969*, Morges, Switzerland, 1970, pp. 19–20

has ever been set up for more than a few days, since the drifting ice is in continuous motion). Central Greenland is the lowest by this standard, with -33 °C. After this come inland stations much farther south, with a continental climate. Thus Oymyakon, at latitude 63° N in eastern Siberia, has an annual mean of -16.1 °C (it also has the lowest absolute temperature anywhere in the northern hemisphere, -67.7 °C). Stations on the coast of the Arctic Ocean are rather warmer: Point Barrow, Alaska, for instance, is -12.2 °C. A distinguishing feature of the Arctic climate, as opposed to that of the sub-Arctic, is cold summers rather than extremely cold winters (see Map 5).

The effect of cold on humans is determined as much by the wind as by the temperature. The concept of wind-chill is useful. Thus a temperature of -15 °C with a wind of 9 m/sec has the same chilling effect as a temperature of -40 °C with a wind of less than 1 m/sec. By this standard, exposed places can be colder than places which are higher in both altitude and latitude. Thus Baker Lake, northwest of Hudson Bay, is physiologically colder in January than Thule in northwest Greenland. High velocity winds, however, are relatively rare in the north; in fact, the low winter temperatures characteristic of Siberia are due to their absence, since only in calm air can the heat loss reach extreme levels.

Precipitation is in general light, with the exception of some coastal regions of the North Pacific and North Atlantic. Snowfall, the main component, was mentioned earlier. Another point worth making is that humidity of the air is also low, because the capacity of cold air to hold water vapour is low. This causes a dryness of skin and especially nostrils which can even become painful. Saturation point is therefore quickly reached, and this causes the appearance of the fog called frost smoke, which forms over open water. When air is calm and below about -35 °C, a continuing source of water vapour will produce ice fog. Such a source may be a car exhaust, industrial combustion products or cooling waters, or just human and animal exhalation. A strong inversion, which is a frequent accompaniment of the required air conditions, will tend to keep the fog in place. So any town or settlement in a sheltered locality is exposed to the risk of this inconvenient and unhealthy hazard. The best-documented example is at Fairbanks, Alaska, but the phenomenon must occur at many other places also, and is likely to expand with northern industrialization, unless preventive measures are taken.

Optical difficulties may be created by the climatic components of the north. One of these is white-out, a condition brought about when a uniformly white surface lies under low cloud, which admits diffuse and shadowless illumination. Surface features and the horizon itself are blotted out, even though particular small objects may still be visible. The effect of this is loss of depth perception – a state which can be very dangerous to aircraft pilots at take-off or landing, and can cause great confusion to surface travellers.

Although the Arctic Circle is a bad indicator of climate, it does indicate the astronomical phenomenon of limited daylight and darkness. These increase regularly, from one day of each on the Circle to six months of each at the Pole. But in fact the period of darkness is less than it might seem to be. Twilight, defined as lasting until the sun is 6° below the horizon, can add several hours of daylight, morning and evening, and refraction further extends the period. A snow-covered surface, moreover, is an excellent reflector, and moonlight and even starlight can provide surprisingly good illumination. From the point of view of effect on human beings and their activities, the total absence of darkness can be disrupting, but in a different way.

Much interest attaches to the question of climatic change. Analysis of mean temperatures at a number of Arctic stations over periods from 1880 to 1960 shows a marked warming which reached its peak in the decade 1930–40. After that, the trend was reversed, but not in a very clear-cut way. The values in the late 1960s were still higher in most places than the mean over the whole period, and it was thought by some Soviet climatologists that the warming period was not necessarily yet over, even though changes in particular months at particular places had been considerable. But later evidence from the worsening floating ice in the seas north of the USSR contradicts this optimistic view. This is a point of some significance for the USSR, where shipping traffic in Arctic waters was developed in the period of warming. Recent deterioration, in so far as it has affected sea ice distribution, has been offset by improvements in ship design and power, and in understanding of sea ice behaviour. It remains a question whether further deterioration could be offset by further technical improvements.

These remarks have been deliberately confined to a few general topics in the climatology of the Arctic and sub-Arctic. In the last twenty years large advances have been made in knowledge of the meteorological processes which occur there. These advances have been made possible by the great increase in data collection brought about by the establishment of weather stations on land and of drifting stations on the Arctic Ocean ice, by observation from aircraft, and by use of more sophisticated equipment. The results have been summarized by climatologists, and it would be out of place to examine them here.

Flora and fauna

The northlands have more abundant plant and animal life than might be supposed. No one seeing for the first time the tundra in flower could fail to be surprised – especially if he had seen it in winter. Two factors which are not obvious at first glance contribute to this: the summer, though short, provides a very intensive growing season because darkness is entirely absent; and although the frost-free period is even shorter, the closely packed and dark-coloured tufts of vegetation can create a microclimate which may be as much as 20 °C warmer than the air above. Temperature is not, in fact, the greatest obstacle to plant growth. Other hazards are wind-chill, scanty precipitation, and low fertility of the soil, which in many parts of the north has emerged from under the great Quaternary ice sheets only in the geologically recent past. Nevertheless much does grow: in the tundra there are grasses, sedges, lichens, and dwarf willows; in the taiga, or boreal forest, conifers, birches, and aspens. All this is the basis for a considerable animal life.

A characteristic feature of the northern flora and fauna is that the number of species is relatively small, while the number of individuals in a species is relatively large. This in turn makes for instability in the ecosystem, since the number of inter-relationships is limited, and the dependence of any single element on the others is increased. Thus populations may fluctuate dramatically, and this frequently occurs in cyclical patterns. A significant delay in the melting of a winter's snow may cause large numbers of birds to fail to find nesting sites. The consequent drop in numbers of young birds allows a massive increase in the insect population, and also obliges weasels and foxes to prey more heavily on lemmings. The imbalances thus caused may last for several seasons.

Land animals which remain the year round in the high Arctic include polar bear (*Thalarctos maritimus*), which spends most of its life on the floating ice, muskox (*Ovibos moschatus*), Arctic fox (*Alopex lagopus*), lemming (*Lemmus* and *Dicrostonyx* sp.), snowy owl (*Nyctea scandiaca*), raven (*Corvus corax*), and ptarmigan (*Lagopus lagopus*). Others, such as the reindeer (*Rangifer tarandus*; called caribou in North America), go south in the winter, into the forest. In the forest the year round are elk (*Alces alces*; called moose in North America), bear (*Ursus* sp.), and many small fur-bearers. Of the birds, which are very numerous in the northlands in summer, most migrate to the south in winter – the Arctic tern (*Sterna paradisea*) going as far as Antarctica, a round trip of 30,000 km. There are few amphibians in the northlands, and no reptiles or marsupials.

The northern seas support very abundant life. In regions where the floating ice gives way to open water, and warmer Atlantic or Pacific waters mix with those of the Arctic, biological productivity is at a high level. Large populations of cod, halibut, salmon, char, shrimp, and crab are found, and all are fished commercially. A notable feature of northern seas is the relative abundance of sea mammals. Among the seals are several species of true seal (*Phocidae*), and the valuable northern fur seal

(*Callorhinus ursinus*), which breeds at only three places in the North Pacific. Both Atlantic and Pacific walrus are found (*Odobenus rosmarus*), and several species of whale, including the killer whale (*Orcinus orca*), the white whale or beluga (*Delphinapterus leucas*), and the narwhal (*Monodon monoceros*). Many of these are found in areas in which ice is present, but only a few are able to remain in winter and solve the problem of finding breathing-holes in continuous ice cover. Such ice cover inhibits most forms of life, and for this reason the central parts of the Arctic Ocean are the poorest biologically of any oceanic area in the world.

The northlands, then, apart from the small area covered by land ice, are by no means barren wastes. But if one looks at their biological productivity from the point of view of human settlement and its associated economic problems, one must conclude that the region is not likely to provide food for the rest of the world. The exception to this conclusion is in fisheries, which do now, and probably will increasingly, make a substantial contribution. The land flora and fauna have up to now been able to support only a very sparse native population, which as it grows in size, is already exceeding the capacity to produce food in several parts of the north. But this does not mean that the capacity cannot be increased. Animal husbandry can be introduced or expanded. Reindeer farming takes place throughout the northlands of the eastern hemisphere, and could be much expanded in the western. Cows, horses, and sheep have been successful in some areas and could be in many more. The growing of cereal crops and vegetables is possible in many places, but has not generally been economic. The natural environment in the north is clearly capable of producing more food. But this will be linked to local needs.

Human pressure on the biological resources of the north has already led to many changes in the balance of nature, to the disappearance of some species, and to a threat to others. On land, there has been a dramatic drop in caribou numbers in Canada, from 3 million to 200,000 in thirty years; and a fall of comparable dimensions in the fur-bearing sable (*Martes zibellina*) of northern Siberia was arrested only by protective legislation during the present century. The muskox was threatened with extinction in northern Canada at the hands of exploring expeditions in the nineteenth and early twentieth centuries, and is now protected. Of marine life, Steller's sea cow (*Hydrodamalis gigas*) became extinct, presumably through overhunting, before the end of the eighteenth century. The northern fur seal and sea otter (*Enhydra lutris*) might have gone the same way had they not been strictly protected. Today, the polar bear and the walrus are both threatened, and are protected by most northern powers. Agreement has been reached between the USSR, Norway and Canada on conservation of the harp seal (*Pagophilus groenlandicus*), much hunted in the North Atlantic. There have been cases, too, of the introduction of new species. Thus the domesticated reindeer was introduced from Siberia into Alaska in 1892 and from Alaska into Canada in 1932, the muskrat (*Ondatra zibethica*) from Canada into Siberia in 1927–32, and pink and chum salmon (*Oncorhynchus gorbuscha* and *O. keta*) from the Pacific into the Barents Sea in a series of introductions starting in the 1930s and continuing in 1972. Not all such introductions, however, have been successful.

CHAPTER TWO

Northern USSR: the north in a socialist economy

Geographical background

The northlands of the USSR constitute by far the biggest national unit within the region with which this book is concerned. The Arctic Circle runs through Soviet territory for over 160° out of 360°, and the most northerly point of the Eurasian mainland (Mys Chelyuskina, lat. 77° 43′ N) is 630 km nearer the Pole than its North American equivalent (the northern tip of Boothia Peninsula, lat. 72 °N). There is no generally accepted southern limit of 'the north', but definitions put forward by S. V. Slavin of Gosplan, the state planning commission, are gaining acceptance. He defines an area he calls 'the Soviet north' in terms of economic development, selecting population density of less than 5 persons per sq km as the main criterion. This gives a southern limit at about latitude 60° N in Europe, dropping to 50° N at the Pacific seaboard. The area of the Soviet north, so defined, is about 11.1 million sq km, or half the land area of the USSR. This is substantially bigger than the whole United States, including Alaska (see Map 6).

This immense northern territory is, of course, diverse in character. The western half is predominantly flat, being the northward extension of the Russian plain and of the west Siberian plain. It is intersected by the Ural mountains which separate the two and rise to heights of between 1,000 and 2,000 m. The eastern half is mountainous. The central Siberian plateau, 400 to 1,700 m high, occupies the region between the Yenisey and Lena rivers, while east of the Lena a series of great ranges, with heights

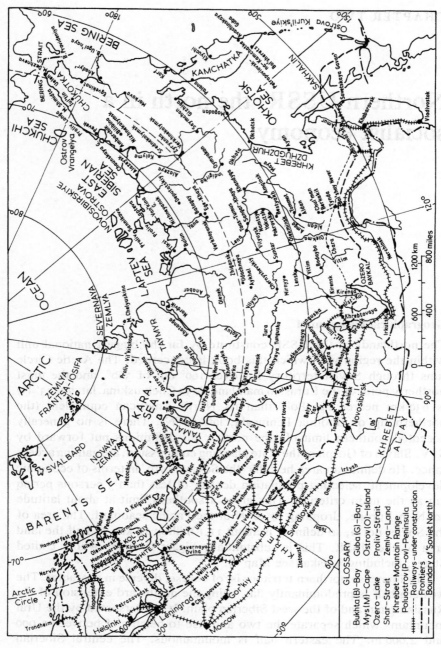

Map 6. The Soviet north

up to 3,000 m and sometimes ice covered, extend to the shores of the Okhotsk and Bering Seas. The highest mountains of all are found in the southern appendage of this region, the Kamchatka peninsula, where volcanic peaks exceed 4,000 m. The long Arctic coast is much more open than its North American analogue, but there are several important island groups lying offshore. The coastal seas – Barents, Kara, Laptev, East Siberian, and Chukchi – are mainly on the continental shelf and therefore shallow.

By far the greater part of this whole area is forested. The boreal forest, or taiga, stretches from the Finnish frontier to the Pacific. West of the Urals it is predominantly pine or spruce. To the east, larch predominates, but the forest is thinned by the marshland of the west Siberian plain and by the altitude of the mountains beyond the Lena. In the north, the forest gives way to tundra between latitudes 66° and 70° N, but in the south the change to steppe takes place beyond our southern limit.

In the tundra, the fauna has relatively few species, but sometimes very large numbers of each: reindeer, both wild and domestic (wild reindeer are the caribou of North America), Arctic fox, lemming, ptarmigan, snowy owl, snow bunting, various ducks and geese, several salmon species, whitefish, and char. In the forest there are elk, reindeer, bear, squirrel, hare, fox, marten, ermine, capercailzie and many other birds, and most of the same fishes as in the tundra.

The river drainage is almost all northwards. Three Siberian rivers – the Ob', Yenisey, and Lena – are among the ten largest rivers of the world, and so is the Amur, the one major exception to the south–north orientation. Even lesser rivers such as the Pechora, Indigirka, and Kolyma are large by ordinary standards. All the rivers are icebound for part of the year, so the fact that the headwaters of most are in the south leads to extensive annual flooding as the spring melt progresses northwards. Owing to the high water levels much timber is carried out to sea, and provides driftwood on the shores of the treeless islands of the Arctic Ocean and Greenland.

The most important conditioning factor, from the point of view of human activity as well as of the rest of nature, is the climate. It is everywhere more severe than in the developed areas to the south. In the islands and the tundra (whose southern boundary, the tree line, roughly coincides with the 10 °C July isotherm), cool, short summers rather than very cold winters are the distinguishing feature. The proximity of the sea and of the air masses associated with it bring about relatively moderate winters, with temperatures seldom falling below −40 °C. Mildest of all is the western end, close to the Norwegian frontier, where the North Atlantic drift makes an ice-free area highly useful to the USSR. In the forest zone, south of the tundra, continentality causes greater extremes of temperature. The western half, west of the Urals, is again the milder. In the east, the dominant feature is a strong anticyclone lasting from October to March, and this causes cold, sunny, and dry winters. Mean January temperatures of -50 °C are found in some places, and absolute minima of nearly -70 °C are recorded for Verkhoyansk and Oymyakon – the latter a frost hollow 350 km south of the Arctic Circle. Summer temperatures of over 30 °C are common, so the annual amplitude in this region can exceed 100 °C – the highest in the world. Winds are relatively weak, and precipitation is light, not exceeding 60 cm of water equivalent annually west of the Urals, and 35 cm east of them.

An important result of past climate in the region is the presence of permafrost. In European Russia it underlies a relatively narrow coastal strip north of the Arctic Circle, and in western Siberia a somewhat broader one; but east of the Yenisey the southern boundary bends abruptly southwards to enclose almost the whole country from there to the Pacific. Permafrost underlies 47 per cent of the territory of the USSR. The frozen layer is often of the order of hundreds of metres deep, and sometimes over a thousand. It becomes thinner and then sporadic as latitude decreases. The presence of permafrost has a very important influence on construction works of all kinds, and failure to take this into account can lead to serious trouble. Its presence may also have a profound effect on vegetation; in areas of low precipitation forest flourishes only because the permafrost, being impermeable, holds the moisture near the surface.

History

The interior of northern Eurasia was explored almost entirely by Russians, or by persons in the Russian service. Fisherman and peasant moved northwards into the White Sea region and the Pechora basin from about the eleventh century, and fur traders and cossacks first penetrated the region east of the Urals in the sixteenth and seventeenth centuries. There was a spectacular advance across Siberia, from the Urals to the Pacific, between 1580 and 1640 – 5,000 km in sixty years. The motivation for the great eastwards movement was the search for fur, and the speed of advance was to some extent brought about by the need to find fresh hunting grounds after the fur-bearers had been severely reduced in numbers by the initial onslaught. The lands the Russians occupied were inhabited by hunters, fishers, and reindeer herders belonging to fifteen or more distinct national groups, yet the Russian conquest of north Asia was accomplished more by a process of infiltration than by military action. Some battles were fought, but not very many, and most involved only a few hundred men. Furthermore, the natives lacked firearms, which the Russians possessed.

The routes used by the incoming Russians were the waterways, which could be travelled by boat in summer or by sledge in winter. The northward-flowing rivers of Siberia have tributaries which run east–west and thus form good routes across the country. When the first wave of fur-hunters and traders had come through, there was not a great deal of consolidation, but there was enough to preserve a Russian presence throughout the region. The Pacific seaboard was the last part to be reached by colonists. Kamchatka was annexed by a party under the cossack Atlasov in 1697. Chukotka, the peninsula facing Bering Strait, was the last to be subdued, owing to the warlike nature of the Chukchi. Although Dezhnev had spent several years there in the middle of the seventeenth century, and others had followed him, the Chukchi did not formally become Russian subjects until 1789.

The eighteenth century, in Russia as elsewhere, saw the advent of scientific study and the beginning of exploration for its own sake. The first and second Kamchatka expeditions (1725–30 and 1733–43) under Vitus Bering played a major part. The second included an 'academic detachment' which collected material about inland regions of northern Siberia. Other scholar-explorers followed. In these explorations a significant role was played by persons of non-Russian origin: first by scientists brought into the country as a result of Peter the Great's initiative – Müller, Gmelin, Steller, Delisle, Pallas – and later by Polish exiles such as Piekarski, Czekanowski, Czerski, and Dybowski. The German traveller Middendorff and the Finnish linguist Castren made especially valuable contributions in the nineteenth century. Of course, Russians also played an important part: Krasheninnikov, Zuyev, Sarychev, Kropotkin, and many others were outstanding. A focus for much of this interest was provided by the Imperial Russian Geographical Society, which started a branch in Irkutsk in 1851 and another in Omsk in 1877.

There was never at any time a direct threat to Russian power in northern Eurasia by another country capable of taking and occupying large parts of the territory. It is true that the northern frontier with Scandinavia might have been drawn somewhat differently if Sweden had prevailed in the seventeenth and eighteenth centuries; and that the Russian presence on the Pacific seaboard was probably arousing the suspicion and jealousy of the British and French in the eighteenth century, of the Americans in the nineteenth, and of the Americans and the Japanese in the twentieth. These threats affected only the extreme west and extreme east of the territory, and they were not very serious. A threat which was a real factor from the seventeenth century onwards was rivalry with China, but that never affected the northern regions with which we are concerned in this book. Therefore, Russia was able to retain control of this vast expanse of land without special difficulty, despite a very sparse immigrant population which also contained a significant fraction of unwilling and therefore unreliable elements, and despite a notoriously corrupt and inefficient administration.

Resources and their development

Minerals

In the last fifty years, the main thing that the Soviet government has been seeking in the north has been mineral resources. The need, rendered urgent in Soviet eyes by the almost universal hostility towards the young Soviet state, was to discover and exploit deposits of those minerals in which the country was then deficient. This motivation was strong enough to overcome the disincentive of high development costs. But even so, the emphasis has naturally tended to be on minerals of exceptional value; gold, nickel, tin, diamonds and apatite are the principal ones. Unfortunately, almost all metal output has long been classified information in the USSR, so statistics cannot usually be quoted.

Of these mentioned, *gold* was the only mineral resource exploited before 1917. There were important workings of alluvial gold in the Yenisey and Lena valleys from the middle of the nineteenth century. In Soviet times some of these had already been worked out, but they were replaced by many others. In recent years there has been a tendency to concentrate on the northeast, where many rich deposits have been found. The industry is organised in gold-mining trusts. There have been mines on the Kolyma since 1928 and in the Bilibino region of Chukotka since 1933, and in the Pevek region farther east since 1958. There has been recent expansion farther east still, near Mys Shmidta. All these are the responsibility of the northeastern trust (Severovostokzoloto), centred on Magadan. Production costs here are said to be the lowest in the country. The trust for Yakutskaya ASSR (Yakutzoloto) manages four main groups: on the Aldan, the trust's headquarters and the site of the initial Soviet 'gold rush' in 1923; in the Dzhugdzhur mountains, where mining started in 1933; on the Indigirka from 1940; and on the Yana from 1958. Finally, on the southern edge of the Soviet north, the Lena trust (Lenzoloto) at Kirensk operates mines on the Vitim and Olekma rivers, tributaries of the Lena. This trust boasts possession of the world's largest dredge, weighing 10,300 tonnes and having scoops of 600-litre capacity. This method, by which dredges work alluvium in surface deposits, is the most widely used in the northern areas. There is also a small gold-mining development which has started recently in Kamchatka.

The best published estimates of production worked out by Western specialists give the 1970 output as 98 tonnes for Severovostokzoloto, 49 for Yakutzoloto, and 15 for Lenzoloto, totalling 162 tonnes out of an estimate for the whole country of 347. Of that national total, 68 tonnes are estimated to be obtained either from private producers ('artels') or as a by-product from mining other metals, and some of this also relates to the north. Rougher estimates for 1973 give those three trusts 179 tonnes out of 398 – almost the same proportion. Thus the north accounts for nearly half. The USSR, with probably about a seventh of world output, is the second biggest producer in the world, but a long way behind South Africa, which produces four or five times as much. About three-quarters of Soviet production is thought to go into reserve for foreign currency purchase, and the other quarter is for industrial use, for jewelry, for medical uses, and for sale to other members of the Soviet bloc. The industry seems to have been subsidized at a generous rate, at least until recently, since the industry is believed to have sold to the State Bank in 1970 at 2 roubles per gramme (roughly $70 per ounce) while the State Bank sold at 1 rouble per gramme. The dramatic rise in the price of gold since 1970, however, has no doubt changed this. With a world price of $150 per ounce (1977), the Soviet industry should be making a handsome profit.

Nickel is mined at two sites in the Soviet north, where copper-nickel sulphide ores are found. The first is at Monchegorsk, south of Murmansk, where mining and smelting started at the Severonikel' combine in 1936. The war caused temporary closure, but when it was over the USSR possessed in addition the nickel mine near Petsamo in former Finnish territory, and started to operate it in 1946 under the name of the Pechenganikel' combine. Monchegorsk has now been virtually worked out, with only poor ore remaining, and its production was replaced by that from the former Finnish site, where the town of 21,000 is now named Nikel'. A new deposit called Zhdanov was found near Nikel' and the town of Zapolyarnyy arose there in 1955, and grew to 22,000 by 1970. Monchegorsk, however, remains the principal processing centre, with a population of 46,000 in 1970.

A larger development started slightly later, in 1940, at Noril'sk near the lower Yenisey. The ore body was rather similar, and not only nickel, but copper, cobalt, platinum, selenium, tellurium and other metals are produced here. In 1971 Noril'sk became the biggest copper producer in the country. Communication with the outside world is by air, or by the 96-km railway to Dudinka on the Yenisey and thence either by river upstream to Krasnoyarsk, or by sea to Murmansk or Arkhangel'sk. The original ore body was evidently large, but a second one, even larger, was found in 1961 at Talnakh, 23 km to the northeast. It is now being developed as the original sources become depleted. Four mines have been sunk there: Mayak (opened 1966), Komsomol (1971), Oktyabr' (1974) and Taymyr (started 1973). Oktyabr' is due to become the largest producer of non-ferrous metals in the USSR when it reaches its designed capacity in 1980, and investment in it is said to be comparable to investment in the Bratsk hydroelectric station, which was the biggest in the world when it was built. There is a smelter at Noril'sk, and methods powered by both electricity and coal have been used. A coking coal is found locally, and up to 3 million tonnes a year have been mined. But electrical methods are preferred and, with the completion of the hydroelectric station at Khantayka (see p. 37) and the planning of another on the Kureyka, will no doubt prevail. The arrival of a gas pipeline in 1969, followed by a second in 1973, further eased the energy situation. A second smelter, to deal with the increased output from Talnakh, was due to come into use in 1975, but was probably completed only in 1977. The profitability of the whole Noril'sk mining complex is reported to be very great.

The annual output of nickel from these two groups of Arctic deposits – the Kola region and Noril'sk – was estimated by an American source to be 135,000 tonnes in 1973, and may therefore be about two-thirds of total Soviet nickel production. The Soviet total is in turn about half the Canadian total, which is the largest in the world. Expansion of mining at Noril'sk is evidently proceeding faster than expansion of smelting facilities there. Ore is now sent from Noril'sk, by the northern sea route, to the smelter at Monchegorsk, which serves both Severonikel' and Pechenganikel', and where there is surplus capacity.

The population of Noril'sk and of the surrounding small towns (Kayerkan, Valek, Ugol'nyy, Medvezhiy, Talnakh) was well over 150,000 in 1970, and so must have been at least 180,000 in 1976, when Noril'sk itself had 168,000. It is the biggest city north of the Arctic Circle in Asia. There is secondary industry, and Noril'sk is a scientific and cultural centre. Interestingly, it is not an administrative centre by deliberate policy. The capital of the Taymyrskiy NO, the territory of which surrounds Noril'sk, is Dudinka, a tenth of its size. Noril'sk itself comes directly under the Kray administration at Krasnoyarsk. A measure of its importance in the Soviet economy is the fact that the director of the mining and metallurgical combine there, V. I. Dolgikh, left to become Party Secretary for the Kray and then in 1972 a Secretary of the Central Committee and as such one of the twenty or so most powerful men in the country.

Tin is another major product of the Soviet north. In the early 1940s mining was started at Ege-Khaya (also spelt Ese-Khaya), in Yakutskaya ASSR, and near Pevek in Chukotka. Some tin was also obtained from the gold-mining region of the Kolyma, but all mines here, except for one at Omsukchan, were closed by 1957. The loss was replaced by two other Arctic mining centres: Deputatskiy, northeast of Ege-Khaya, where placer mining started in 1953, and Iul'tin, southeast of Pevek, which opened in 1959. Both are in remote locations which pose difficult transport problems. Deputatskiy has been linked by winter road to the Magadan highway, while Iul'tin is connected by road to the specially enlarged port of Egvekinot on the Bering Sea. The concentrate from both areas is believed to be taken to the smelter at Novosibirsk. Offshore dredging for tin started at Van'kina Guba, on the Laptev Sea, in 1974. Soviet sources state that these Arctic deposits produce about half the total Soviet production of tin, which was estimated by American specialists to be 39,000 tonnes in 1973. The country is not yet self-sufficient in this respect, and has to import.

The development of a *diamond* industry in central and northern Yakutskaya ASSR has been a major feature of the northern industrial program. The country was almost entirely dependent on imports until production here started in 1957. Kimberlite pipes were found on a tributary of the Vilyuy, where the centre for the industry, Mirnyy, was built; and also 400 km to the north, where settlements named Aykhal and Udachnaya were created. These three are worked – Mirnyy fully, the other two probably not yet fully; and many other pipes have been found in this general area of northwest Yakutskaya ASSR. The ore-crushing mill is at Mirnyy, and a hydroelectric station has been built on the Vilyuy at Chernyshevskiy, 100 km away. The population of the region in 1971 was 47,000, and Mirnyy itself was 26,000 in 1973. Surface communication with the outside world is chiefly by road to Lensk on the Lena (250 km) and thence by river. Output is not known, but one Western guess was 10.5 million carats in 1969 and 12 million in 1970, and another was 9.5 million carats in 1973. The USSR is believed to be self-sufficient, and she also exports diamonds, using the International Diamond Syndicate to do so. Diamonds are thought to be one of the three major commodities earning foreign exchange for the USSR, the other two being oil and gold. Four-fifths of the diamond output is industrial stones.

Mica (used chiefly in the electrical industry) is obtained from three sites in the Soviet north. Muscovite, which is the rock yielding sheet mica, is found in northern Karel'skaya ASSR, near Chupa and Loukhi, and in Murmanskaya Oblast' at Yena. Near Yena, at Kovdor, there is a deposit of phlogopite, another type of mica. These locations are on the Murmansk railway, which enhances their value (the Yena deposit is of poor quality). The third area is in the Tommot region of southern Yakutskaya ASSR, where the country's largest deposit of phlogopite is found. Mining started here in 1942, and by 1964 four out of nine known sites were being exploited. The sites are spurs of the all-weather road to Yakutsk from the Trans-Siberian railway. No indication of output is available. In all cases, the mineral is concentrated on the spot but sent south for refining. It was reported in 1975 that the Kovdor deposit produced over half the Soviet output of phlogopite.

Tungsten (or wolfram, as it is called in the USSR) is mined in the northeast. There was a mine near Oymyakon, but it was exhausted in 1956. Tungsten is one of the by-products of tin mining at Iul'tin, which has been in operation since 1959.

Among the smaller mining developments is production of the mineral loparite, from which comes niobium (or columbium) and tantalum, important in alloys, at Revda, south of Murmansk. This development started in 1951. A number of other useful minerals are obtained as by-products of mining already mentioned, notably of nickel. They serve as a way of reducing the unit cost of the main operation.

While the foregoing are relatively high-value minerals, the Soviet north also produces low-value minerals, which are nevertheless worth exploiting because of the richness of the ore. One of these is *apatite*, found in great quantity in the mountains south of Murmansk. It is a calcium phosphate which can readily be made into mineral fertilizer, and it has become the principal raw material for the Soviet superphosphate industry. Mining started in 1930. Towns sprang up at Kirovsk (38,000 in 1970) and later at Apatity (56,000 in 1976). Each has a concentrator, which increases the proportion of phosphorus pentoxide from 20–25 per cent in the ore to 40 per cent in the concentrate. Output has steadily increased, and the two concentrators produced 15.4 million tonnes of concentrate in 1975, from 38.1 million tonnes of ore. Transportation is by rail (Apatity is on the Murmansk railway, and Kirovsk is on a short spur of 20 km which was opened in 1930). About half the output is exported, chiefly through Murmansk, while the other half provides three-quarters of the raw material of the USSR's superphosphate industry. A new ore body has been discovered in the region, and production plant with a further capacity of 10 million tonnes a year was planned in 1974. Associated with the apatite is nepheline (or nephelite), which is used as a source of aluminium, and also in glass manufacture. The ore contains 12–14 per cent alumina, and is beneficiated on the site to a 25 per cent concentrate which is sent south. This source is reported to acount for 10 per cent of Soviet aluminium production in 1974.

Another low-value mineral, which is nevertheless exploited, is *iron*. The Kola region is again one of the major sites. Mining started at Olenegorsk in 1955 and at Kovdor in 1962, and the two types of ore are to some extent complementary. Iron content is about 30 per cent, and the ore is concentrated to 62 per cent. Output in 1975 was 14 million tonnes from Olenegorsk and 9 million tonnes from Kovdor, jointly yielding 9.5 million tonnes of concentrate. This goes chiefly to the Cherepovets steel mill, but some (1.8 million tonnes in 1969) is exported. Olenegorsk (24,000 inhabitants in 1974) is on the Murmansk railway, but a 100-km spur had to be built to Kovdor (14,000 in 1970). The two mines account for just under 4 per cent of Soviet iron-ore production. Interest is being shown in the Kostamuksha deposit in western Karelia, where the ore is also only 30 per cent, but is 350 km closer to Cherepovets, and can be concentrated by a cheaper method than is used in the more northerly deposits. The planned capacity of the plant is 24 million tonnes of ore, producing 8 million tonnes of 65 per cent concentrate, and production may start in 1978. An 80-km railway will have to be built.

There are two other northern iron deposits of great potential but neither is yet being worked: the Angara-Pit basin, near the Angara-Yenisey confluence, where the ore is 40 per cent iron and 70 per cent of it could be surface-mined (the Angara-Ilim basin, where mining has started, is south of our area, near Bratsk); and the region of Chul'man in southern Yakutskaya ASSR, where the iron content runs to over 50 per cent, and where, most important of all, there are huge reserves of coking coal within 100 km. The importance of developing the Chul'man iron and coal to form a metallurgical base for the Soviet Far East has often been mooted, and it is fair to assume that this would have a higher priority than the Angara-Pit basin. A railway is now under construction (see p. 67).

Mineral deposits which are known but not worked have not been mentioned in this section. They are very numerous. Two copper deposits deserve mention. It was announced in 1974 that the country's biggest native copper ore deposit was in Taymyr (exact location unspecified) and had recently been prospected. The ore lies at shallow depth and the copper is remarkably pure. The other is the long-known Udokan deposit northeast of Baykal.

Fuels

There is only one *coal* deposit of national significance in the Soviet north, and that is the bituminous coking coal of the Pechora coalfield around Vorkuta, in the northeastern corner of European USSR. It was first exploited just before the Second World War, but became important only when the railway linking it with the rest of the country was completed in 1942. At that time the coal was desperately needed to replace output from the overrun Donbass. But this need disappeared relatively soon, and the coal was then used to supply Leningrad and other northern centres, as originally intended. Production has risen steadily but rather slowly, reaching 24.2 million tonnes in 1975, 70 per cent of it coking coal. Further expansion is taking place, and the Vorgashorskaya-1 mine, brought into production in 1975, has an output capacity of 4.5 million tonnes a year – the largest in Europe. The cost per tonne mined at Vorkuta is above the national average, but reserves are bigger even than those of the Donbass. The population of the coalfield region was about 200,000 in 1965, and Vorkuta itself 96,000 in 1976.

Coal is mined at many other points, but for local use only. The remoteness of many northern settlements makes worthwhile the exploitation of almost any local source of fuel, however poor the quality. Among the deposits worked are those of Noril'sk (in fact a high quality coal, used in the nickel-refining process); in the Lena basin, particularly at Sangar-Khaya and Dzhebariki-Khaya, and at Sogo, near Tiksi; on the Kolyma, at Zyryanka; and on the shore of the Bering Sea at Bukhta Ugol'naya.

Another major development will be the exploitation of the southern Yakutskaya ASSR coal in the Chul'man region. This has been mentioned in connection with the nearby iron deposits, and would have national significance. There has been local mining here since the 1930s, and output in 1965 exceeded 200,000 tonnes. The main mining town, Neryungri, became a town in 1975, implying increased importance. But major development will come with the railway (see p. 67). This was speeded by signature in 1974 of an agreement with Japan. Japan will provide credits, to be used largely to build a railway, and in return is to receive 5 million tonnes of coal a year, starting in 1983 and continuing until the total reaches 84 million tonnes.

The total coal production in the 'Soviet north' in 1974 was 35 million tonnes, two-thirds of it from the Pechora coalfield.

Of much greater importance on the national level are the northern sources of *oil and gas*. The major area of discovery is the northern part of the west Siberian plain, between the rivers Ob' and Yenisey. This is now known to be one of the great oil and gas provinces in the world. Gas was found at Berezovo on the lower Ob' in 1953, and exploitation started there in 1963. Oil was first found in 1960 farther upstream, on the middle Ob', in the region of Ust' Balyk (later called Nefteyugansk) and Megion, and was exploited from 1964. Other oil and gas fields were found on the Vasyugan, a left tributary of the middle Ob'; on the peninsula of Yamal; and in the region of the rivers Pur and Taz. These points are within a rectangle about 800 km from east to west and 1,200 km from north to south. The biggest oil reserves are in the south, on the middle Ob', and the biggest gas reserves in the north, in the Pur basin. The petroliferous province is thought to extend farther in all directions, especially northwards into the Kara Sea. The main centres are the rapidly growing town of Surgut (67,000 in 1976) and the new towns of Nefteyugansk (31,000 in 1973) and Nizhnevartovsk (63,000 in 1976), all in the oil area, and Nadym (15,000 in 1972), the main gas centre. The number of fields and size of the reserves is continually being upgraded as exploration continues. In summary, in 1974 there were 245 proved fields of oil, gas, or condensate (Table 1). The age was mostly Jurassic or Cretaceous (Mesozoic), and the depth between 2,000 and 3,000 m.

None of these fields are near potential users. Exploitation therefore depends on solution to transport problems. Oil from the middle Ob' was at first moved by water; but this method, being seasonal, could never be more than a temporary solution, and pipelines have always been seen as the right answer. A very extensive network has been planned, and a

Table 1. Oil- and gasfields in the west Siberian plain, 1974

Petroliferous division and subdivision	Date of first discovery of commercial quantity	No. of proved fields in subdivision	Oil	Conden-sate	Gas
Ural region					
Berezovo (incl. Punga)	1953	21		*	*
Shaim	1960	31	*		*
Karabash	1964	1			*
Frolov region					
Kazym	1963	3		*	*
Krasnoleninsk	1962	6	*		*
Tobol'sk	1971	1		*	*
Kaymysovy region					
Dem'yansk	1964	5	*		
Kaymysovy	1964	9	*		*
Mezhov	1962	9	*	*	*
Vasyugan region					
Pudinsk	1967	6	*	*	*
Vasyugan	1962	14	*	*	*
Aleksandrovskoye	1962	12	*		*
Payduginsk region					
Sil'ginskoye	1961	6	*	*	*
Payduginsk	1969	1	*		
Middle Ob' region					
Salym	1964	8	*		*
Surgut	1961	34	*	*	*
Nizhnevartovskoye (incl. Samotlor)	1961	32	*	*	*
Nadym-Pur region					
Purpey	1965	13	*	*	*
Nadym (incl. Medvezh'ye)	1966	2		*	*
Urengoy	1966	4	*	*	*
Gyda region					
Yamburg	1969	3		*	*
Srednemessoyakha	1971	1			*
Tambey	1973	1			*
South Yamal region					
Nurminsk	1964	8	*	*	*
Pur-Taz region					
Taz	1962	4	*	*	*
Sidorov	1972	1			*
Ust' Yeniseysk region					
Tanama	1967	7			*
Dorofeyev	1966	2			*
TOTAL		245			

(*Source*: A. E. Kantorovich *et al.*, *Geologiya; nefti i gaza Zapadnoy Sibiri*, Moscow, 1975, pp. 432–52. Chemical analysis of the oil in each field is given in Z. V. Driatskaya and G. Kh. Khodzhayev (eds), *Nefti SSSR*, Tom 4, 1974, pp. 381–628.)

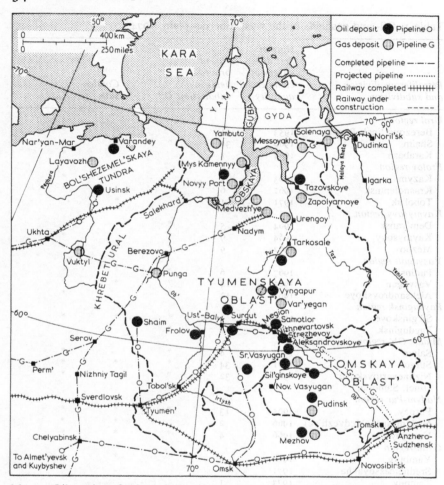

Map 7. Oil- and gasfields and pipelines in the west Siberian plain, 1975

considerable part of it has been built (see Map 7). The pipelines are either 1.42, 1.22 or 1.02 m (56, 48 or 40 in.) in diameter. Most of the oil goes to European USSR for refinery, but some is refined at Omsk and at Angarsk, near Irkutsk. Two more Siberian refineries, at Achinsk and Pavlodar, are due for completion in 1978.

Casinghead gas is processed at Nizhnevartovsk, and supplied to the major thermal power station for the area at Surgut. Natural gas condensate is processed at Dudinka to produce petrol for local consumption. Siberian gas reached Moscow in 1975. It came from the Medvezh'ye field, and followed the more southerly Urals, crossing via Serov. The original plan for a pipeline crossing of the Urals near Salekhard was dropped, but another crossing, in a mid-way position, from Punga to Vuktyl', was decided upon and started in 1975. This pipeline is designed to take Siberian gas to the west by way of the so-called 'Northern Lights' pipeline (*Siyaniye severa*) as the Vuktyl' reserves run down. Double and triple stringing of many pipelines has been undertaken. Future development of gas is planned to move from Medvezh'ye eastwards to Urengoy, where commercial production is expected in 1978, and then to the Gubkinskoye, Vyngapur, and Var'yegan fields. The Surgut–Urengoy railway (see p. 66) will come up through these fields and will of course play a key role.

The difficulty of the terrain must be emphasized. The west Siberian plain is one of the largest swamp areas in the world, and there were no roads or railways at all before this development started. Platforms are built, from which wells are drilled at an angle. The southern boundary of discontinuous permafrost runs across the middle of the area, at about the latitude of Berezovo. This has meant that the most severe conditions for construction in or on permafrost have been met by gas pipelines, and gas will still flow at the temperature of the adjacent air or soil. For oil this is not the case, but the main oil-producing region, and the pipelines so far built, are south of the boundary and therefore cross only patches of permafrost. If the difficulty with oil pipelines is the possible melting of permafrost, the equivalent difficulty with gas pipelines is that chilled gas may create permafrost. But although chilling allows greater through-put, no refrigeration has yet been attempted.

The output from the west Siberian region has risen steadily since its start in 1963–4, to an annual level in 1975 of 38 billion cu m of gas (1.34 trillion cu ft) and 147 million tonnes of oil (2.9 million barrels a day). In this context west Siberian means Tyumenskaya Oblast' and Tomskaya Oblast', with the former producing over 95 per cent of the output. Over half of the oil at that date came from one pool, Samotlor. This represents an overfulfilment of the 1975 plan for both oil and gas. The tenth five-year-plan (1976–80) calls for a level in 1980 of 155 billion cu m of gas (5.5 trillion cu ft) and 300 million tonnes of oil (6 million barrels a day). The oil target for 1980 represents an increase of 20 per cent on the target for 1980 issued in 1974. West Siberian production accounted for 30 per cent of the total Soviet output of oil, and 13 per cent of the output of gas. In 1980 these figures should be 50 and 35 per cent respectively. The reserves are already vast and are certain to grow as exploration continues, especially into deeper strata. But the limiting factor in growth is not the size of reserves, but practical problems of moving the output, and administrative problems of financing the whole development. The main problem of the immediate future, as seen in 1975, was that many widely scattered and relatively smaller deposits would be brought into production, thus causing much more difficulty than when almost total reliance was placed on one oil pool (Samotlor). Linked to this are the problems of seasonal working. Construction work has hitherto had to be restricted to the five winter months, because no way of overcoming the bog has been found.

While the west Siberian basin is the most important region for development at present, it is not the only one in the Soviet north. There are three others. The Pechora basin, west of the Urals, has oil and gas, and in 1975 produced 7.1 million tonnes of oil (140,000 barrels a day) and 4 million tonnes of natural gas liquids in the region centred at Ukhta, where there is a refinery able to deal with part of the output; and 18.5 billion cu m of gas (two-thirds of a trillion cu ft), most of it from Vuktyl' and piped to Torzhok, between Leningrad and Moscow, in the 'Northern Lights' pipeline. Development of a new gasfield, Layavozh, east of Nar'yan Mar and of the same magnitude as Vuktyl', started in 1974. A new oilfield centred on Usinsk came into production, behind schedule, about the same time, and should triple Pechora oil production to 22 million tonnes. It is being extended northwards to the coast, where an exploration well struck oil at Varandey in 1975. The Vilyuy basin in Yakutskaya ASSR produces gas at Taas-Tumus, and this is piped to Yakutsk and Bestyakh on the Lena. In 1974 0.4 billion cu m was produced. Production could be greatly increased if a larger market were found. Magadan has been suggested, and there have also been discussions with Japan and the USA. In this context an outlet on the Pacific coast would be required, and studies are now under way. Proved reserves were 220 billion cu m in late 1971, and these were doubled in 1973 by the connection to the pipeline of the recently found Mastakh field, 150 km west of Taas-Tumus. Finally, the Anadyr' basin, on the shore of the Bering Sea, is known to contain gas, but there is no exploitation yet.

An American estimate made in 1971 puts the probable reserves of all four Soviet northern basins at 9 billion tonnes of oil (66 billion barrels), and 21.5 trillion cu m of gas (757 trillion cu ft). What has so far been found in these basins is of course only a fraction of what will be found, and these estimates are already very conservative. The proven gas reserves in Tyumenskaya Oblast' alone by late 1974 were 15 trillion cu m, and the Soviet prediction for that area was 70 trillion cu m. Exploration continues in many areas. The offshore prospects are particularly good, and drilling started in 1972 on Ostrov Kolguyev, in the southeast corner of the Barents Sea. The northern part of the west Siberian province (including offshore) is likely to be the most promising area for the next major exploration effort. Indeed, it is thought that offshore potential on the wide continental shelf along the whole north coast of the USSR might increase the estimates given above by a factor of anything between two and five.

Electric power

There are a large number of small thermal power stations in the Soviet north, associated with local industry. No special interest attaches to most of these. They are of standard designs, and are fuelled by coal, natural gas, oil, or even wood. The major thermal station using casinghead gas, under construction at Surgut, is to have a capacity of 2,400 MW, of which a first stage of 1,200 MW was completed in 1972–5. Of more novel design is the series of so-called 'Northern Lights' (this time *Severnoye siyaniye*) power stations, which are mounted on barges. They are oil-fuelled, produce 20 MW at 6,300 volts, and can be towed into position anywhere on the river system where draught (only 1.55 m) permits. The first model was built at Tyumen' in 1968–9, and has been in use at Zelenyy Mys, below Nizhnekolymsk on the lower Kolyma, since 1970. Three similar models have been built, and are stationed on the Pechora, on the Aldan, and at Mys Shmidta. A fifth was due for completion in 1977, and others will no doubt follow, for they have proved very useful. The idea is basically a variant, designed to meet northern conditions, of the common Soviet practice of mounting generators on rolling stock.

Two nuclear power stations are functioning in the north. One is in the Murmansk region, an area lacking in both fossil fuels and large rivers. The power station is referred to as the Kola nuclear power station and will have four 440 MW generators of which two were producing electricity by December 1974. The other is at Bilibino in Chukotka, to serve the gold-mining industry, and is much smaller (48 MW). Its first stage went into operation in late 1973, and the fourth and last was completed in 1976.

Hydroelectric schemes pose new problems in the northern environment, and therefore present more interest. There are several on rivers south and west of Murmansk, but this is not in the permafrost zone. The most interesting one is at Chernyshevskiy on the river Vilyuy. Here there is continuous permafrost, and this is the material used to form the dam, which has therefore to be kept frozen in order to remain effective. The station's eight generating units, completed in 1976, produce 650 MW, used by the diamond industry. Another important hydroelectric station is that on the Khantayka, a tributary of the Yenisey near Noril'sk, for which the power is needed. Construction started in 1963, and was completed in late 1972. Capacity is 441 MW. A second station to supply Noril'sk is under construction on the Kureyka, a more southerly tributary of the Yenisey. It will have a capacity of 500 MW and should enter service in 1981. Both here and at Chernyshevskiy a town had to be built to house the construction workers. Finally, plans for a 750 MW station on the Kolyma have been approved. Construction started in 1971 and the first stage is to be on load by 1980.

On the southern edge of our area an important chain of very large stations is in course of construction on the Angara river, between Baykal and the Yenisey. The first to be completed was that at Irkutsk (660 MW, 1958) and the second at Bratsk (4,100 MW, 1966). The third station, Ust'-Ilimsk, has a designed capacity of 3,600 MW, and the first three sets (out of fifteen), each generating 240 MW, entered service in 1974, and completion was expected by late 1977. The fourth station is to be Boguchany (4,000 MW); construction started in 1974, and operation should begin in the mid-1980s.

A pilot scheme for a tidal power station was put into operation in 1968 at Kislaya Guba, an inlet near Murmansk. This has a capacity of 1.2 MW, but it is hoped to build larger plants farther east along the coast in the same general area, and at the mouths of the Mezen' and Kuloy rivers, on the other side of the White Sea. The most ambitious project of all calls for a dam across Penzhinskaya Guba, in the Sea of Okhotsk, and would produce at least 2,500 MW.

Some other very ambitious schemes have been suggested. A 20,000 MW station at the mouth of the Lena; another of 5,000 MW, but harder to build, at the mouth of the Ob'; and the famous Bering Strait dam. Of these, the Lena station is mentioned as a real possibility and is said to have been approved in principle (though no one has yet demonstrated the need for such a vast amount of power in so remote a region). The Ob' project seems to have been withdrawn, tacitly and no doubt grudgingly, when it was shown that the very large reservoir would cause a whole series of disadvantageous effects, up to and including flooding most of the wells in the oil- and gasfields. The Bering Strait dam was not to be a producer of electricity but a consumer, since water had to be pumped across the dam. The objective was climatic change over the whole Arctic area. If carried out, it would set off a chain reaction of climatic effects, with unpredictable consequences. The scheme may now be regarded as dead.

Renewable resources

The resource which provided the original motivation for Russians to penetrate northern Siberia was *fur*. In those days (sixteenth to seventeenth centuries) fur was a more widely used and valuable commodity than it is today. Nevertheless, fur remains a significant product of the north.

The introduction of fur farming in the nineteenth century shifted much of the industry away from the north. But some of today's fur farms are located in northern regions, and hunting and trapping in the northern forest does continue. Pelts are obtained from fox, mink, hare, beaver, ermine, sable, and squirrel. The two last-named are more frequently obtained by hunting than by breeding in captivity. The sable, whose pelt is perhaps the most valuable of all, has been the subject of protective legislation at various times in the past. The wild population has now built up again to levels at which controlled hunting is permitted, but fears have recently been voiced in the press that the control is not effective enough.

Both hunting and fur farming in the north are principally organized as part of the work of state and collective farms. There are also some freelance hunters, especially among the native peoples. A new type of institution was introduced in the 1960s, so-called state hunting units (*gospromkhoz*), whose job is to harvest biological resources by hunting, fishing, gathering mushrooms and berries, keeping bees, and so forth. Twenty-five units of this kind were functioning in the north in 1967.

That part of the fur industry located in the north produced in 1966 between 400,000 and 500,000 pelts, 70 per cent of them from fur farms. This was about 10 per cent of the fur production of the whole country, and was worth some 20 million roubles. The Soviet fur trade had a retail turnover of about 950 million roubles in 1974. The total has been steadily rising, due in part to increased use of artificial fur (which is included in the turnover figure). Fur worth 59 million roubles was exported in 1974, and the annual fur auction at Leningrad is still an event of considerable importance to the world's fur dealers, but the relative importance of fur in the country's export trade had been falling. However, northern fur probably constitutes a higher proportion of fur exports than of total fur production, since exported furs are likely to be the rarest and most opulent. The northern industry, therefore, to that extent retains a special importance.

The USSR possesses the largest stand of *timber* in the world, in the boreal forest which stretches across most of the country in a belt 1,500 km deep. The more accessible southern part of this belt is the site of most exploitation, and it lies outside the northlands. The forests within the northlands are not only less accessible and more distant from markets; the trees are also smaller, slower-growing, and of inferior quality. The time needed for spruce to reach a diameter of 24 cm (the normal requirement for structural timber) is anything from 130 years upwards in the northlands, whereas it will be less than 100 farther south. However, the trend since the 1950s has been to increase the cut taken in the remoter, densely forested (*mnogolesnyy*) areas and to decrease that in the closer but more sparsely forested areas; in 1973 the proportion taken from each was about three-quarters and one-quarter respectively. But these densely forested areas, while they include the northlands, also include large areas to the south.

The northlands themselves produced 17 per cent (66 million cu m) of the forest products total for the country in 1974. The main harvested areas in the north are in Arkhangel'skaya Oblast' and in Komi ASSR – the region to the east of the White Sea and in the Pechora basin. Some of the production from this region is exported through Arkhangel'sk. At lower latitudes, exploitation from the southern side of the forest belt has crept northwards at several points. In particular, a railway spur from Ivdel', in the northern Urals, to the Ob' river at Sergino was built in the 1960s specifically to enlarge the area from which timber could be got. Other lines, now under construction or planned, will serve the same purpose, although serving other purposes too (see pp. 66–8). Special importance attaches to Igarka on the lower Yenisey. This is a sawmill centre which draws its timber from higher up the river and exports sawn wood in sea-going vessels, to which the town is accessible. The traffic started in the early 1930s. Since the late 1960s about one million cu m (200,000 standards, weighing about 600,000 tonnes) have been annually exported in this way, in over a hundred ships. This traffic has been growing steadily, for the species made accessible through Igarka (particularly *Larix sibirica*) are of special value. A much smaller development of the same kind has been tried from time to time on the Lena river, the timber being exported to Japan via Bering Strait. The distances involved mean that the operation is likely to be only marginally profitable, and probably this is the reason why it has not grown. A survey of timber prospects in Yakutskaya ASSR published in 1965 concluded that the only parts of the region worth considering for exploitation were those in the extreme south, on the upper Lena and upper Aldan. The situation is therefore not unlike that in the fur industry just described. The northern segment of the timber industry is not very large, but has particular importance because of its ability to earn foreign exchange.

Local *food production* is another renewable resource whose exploitation has some importance. The food produced in the north is not (except in a few cases) marketed outside the north. But its importance lies in the degree to which it can reduce the need to import food from the south.

The most successful development has been that of the *reindeer industry*. In January 1975 there were 2.2 million domesticated reindeer in the USSR, most of them in state farms. The herds are distributed throughout the north, but the greatest density is in the northeast. The total has remained almost static since the early 1960s, and is falling slightly. A native subsistence economy based on reindeer has existed for many centuries. In the money economy introduced in Soviet times the emphasis until the last decade was on use of reindeer for transport rather than on meat production. But wider use of mechanical transport and significant increase in the permitted selling price for reindeer meat made a big difference. In 1967, 722,000 reindeer in Magadanskaya Oblast' yielded 15,000 tonnes of meat (live weight), which would be sufficient for 110,000 persons at the Soviet nutritional norm of 62 kg of dressed meat per person per year (dressed meat may be taken to be half live weight). The population of the region was 352,000 in 1970, so in theory about a third of the meat requirement could be met from local production. In other parts of the north the proportion must be lower, because the density of reindeer is lower. The whole Soviet reindeer industry produced 41,400 tonnes of meat (live weight) in 1971 – sufficient for 330,000 persons or 5 per cent of the population. In practice, however, the meat probably supplies a smaller number of people at a higher level of consumption. Nevertheless, the reduction in the need to import food from the south is significant. From the two meat figures, for the whole country and for Magadanskaya Oblast', it is possible to draw certain deductions. For instance, if 20 per cent of the herd were culled annually (a normal figure for Scandinavia), the average yield of dressed meat per animal would be 43–8 kg, which is higher than in Scandinavia; while a cull of 25 per cent would imply 35–42 kg of meat (normal for Scandinavia). But Siberian reindeer are known to be larger than Scandinavian.

Other domestic animals kept in the north are *cattle and horses*. There were 978,000 head of cattle in the north in 1974, and they are kept chiefly for milk production. The town of Yakutsk, for instance, with a population of over 100,000, is fully supplied with milk by two local state farms. In Magadanskaya Oblast' in 1971 milk production equalled 100 litres per head of population. Horses are found mainly in Yakutskaya ASSR, since Yakuts tend to prefer horsemeat to other meats. Herds totalling 140,000 produced in 1965 about as much meat as the region's 350,000 reindeer, some 7,400 tonnes (live weight). Horsemeat was said to constitute on average 20 per cent of the total meat production of Yakutskaya ASSR in 1961–6 (an unexpectedly large amount of beef is produced here), so the needs of this region are theoretically met from local sources as to about 40 per cent. Analogous figures for other northern regions are not available, but the case of Yakutskaya ASSR is likely to be better than most other regions, with the possible exception of Tyumenskaya Oblast' (the middle and lower Ob' basin), where there are larger numbers of both reindeer and cattle.

The *fish* of northern rivers are an important resource which has long been harvested by the local peoples. The Russians have introduced larger-scale fishing methods and increased the catch. There is no figure available for total northern catch, but an indication may be deduced from results in Khanty-Mansiyskiy NO, the most developed area in this respect. There the catch in the mid-1960s was said to average 20,000 tonnes a year. The whole Soviet north may produce two or three times this amount, which could meet perhaps a third of local demand. However, the fish of Siberian rivers, particularly the salmonate species and the large pike and ide (*Leuciscus idus*), are a delicacy of high value, and therefore are often sent south.

Sea fisheries are not significant in most waters north of the USSR, but they are very important in the Barents Sea. The port of Murmansk (347,000 in 1974), which is at once the largest town within the Arctic Circle, the most northerly port in the country, and the only major one which does not freeze up, is the base for a fishing fleet which ranges over very large oceanic areas. It has been since the 1950s the major Soviet fishing port, handling about a million tonnes of fish in 1970 (more than all British fishing ports put together). Murmansk in the 1950s and early 1960s was a base for the fast-growing Atlantic fishery, but the Baltic ports were found to be more satisfactory in this role, so Murmansk is now more exclusively concerned with the Barents Sea fishery. Arkhangel'sk is also used for this, but to a smaller extent, since it is both more distant from the fishing grounds and is closed by ice for some months each winter. Soviet catches in the Barents Sea have fluctuated with the cod and capelin: from 635,000 tonnes in 1956 to 194,000 in 1958 to 832,000 in 1974. This resource, though originating in the Soviet north, is one which has, of course, national rather than local significance. The port of Petropavlovsk-na-Kamchatke (196,000 in 1975) plays a role on the Pacific analogous to that of Murmansk. The catch passing through this port in 1966 was about 500,000 tonnes, and the figure was growing fast. There is an important king crab industry (*Paralithodes camtschatica*) in the Sea of Okhotsk, off the west coast of Kamchatka.

Crops (other than hay) are not produced on a large scale, despite the impression given in some Soviet literature. There are some model farms and greenhouses, but they do not produce much more than token quantities of food. Of the seven National Districts in the north, which jointly occupy more than half its area, only one, the Khanty-Mansiyskiy, was reported in 1961 to have significant cereal crops and to be self-sufficient in potatoes and vegetables. This not very striking result is the more surprising in view of the apparently effective scientific effort put into developing crops suitable for northern use in the 1920s and 1930s. It seems that the methods, being expensive in time and effort, have not been widely applied. Although the potential is there, improved transport makes it cheaper to bring these foodstuffs in from the south.

Many great *rivers* in the USSR flow northwards. There has been much discussion on the subject of diverting part of the flow southwards, and it now seems that at least some of the action proposed may be taken. A study in depth is in progress by up to a hundred institutes and their reports are due by 1980. The Pechora river is likely to lose some of the flow in its upper half by diversion into tributaries of the Kama, which flows into the Volga. One of the objects here is to try to stabilize the level of the Caspian Sea, which has been falling markedly in recent decades. Work will take ten to twelve years from the start of construction. The relative proximity of the Ob'-Irtysh system to the arid lands of Kazakhstan and

Central Asia has led to the idea of connecting the Irtysh to the Syr-Dar'ya by way of a very long canal through the Turgay valley. This has been suggested at intervals since 1868, but earth-moving equipment has now reached a stage at which the economics of the plan become more thinkable, and some detailed consideration is once again being given to it. A development of this plan, in one of its earlier forms, was that water from the Yenisey should be diverted into the Ob', in order to maintain sufficient flow. The scheme would therefore be very far-reaching. The implications for the Arctic itself in all this are uncertain. The reduction in the flow of fresh water northwards would lead to changes in the ice situation and in fish stocks. But in so far as these are taken into account, they are presumably reckoned a fair price to pay for the irrigation program, which is the motivation for the plan. In 1976 there was a new proposal whereby these two schemes would be combined by taking water from the mouth of the Ob' across the northern extremity of the Urals to the Pechora and thence to the Severnaya Dvina. From there the water would go by existing channels to the Volga, where it would be available both to stabilize the Caspian and, by use of a further canal, to irrigate parts of Central Asia. A scheme as vast as this, if approved, would take many decades to complete.

Towns and settlements

The provision of living accommodation has to be an integral part of any program of resource exploitation in a very sparsely inhabited area. Unsatisfactory housing and limited cultural amenities have always been recognized as two of the main factors acting to dissuade Soviet workers from going north.

Nevertheless, there have been many achievements and many plans. The normal approach in larger centres is to build concrete apartment blocks, up to five storeys high, to a design that seems to be standard throughout the country. Less expensive, but also less commodious, are two-storey wooden houses, generally containing sixteen apartments. These have the advantage that they can be built with local materials by local labour (good carpenters are more readily found in the north than good bricklayers or masons). Few of these wooden houses have drainage, but some have running water. Special efforts have been made in some cases, particularly at Noril'sk and Magadan, where almost all the buildings are of concrete and most have main services. The result is that the population turnover is lower here. On the other hand, they are much more expensive to build.

Many towns in the Soviet north sprawl or straggle over a large area. A strongly made point in recent Soviet writings is the need for compactness, especially near or north of the tree line, where stronger winds create greater problems. Buildings can be sited to protect each other, services can be provided more economically, and pedestrian movement in blizzards and darkness can be made less hazardous.

Plans have been advanced for many sorts of neighbourhood unit embodying these advantages, but it is not clear how many of these have been built. The most advanced idea is for towns with controlled microclimate, where residential blocks would be joined by enclosed streets to an enclosed recreation space under a huge plastic dome. Such towns have been designed to accommodate from 5,000 to 18,000 inhabitants, and were proposed for Aykhal and Deputatskiy, two mining centres in Yakutskaya ASSR. Although the idea was first mooted not later than 1962, neither has been built yet. Meanwhile, a much smaller version, to house 500 gas-pipeline workers at Poluy in northwestern Siberia, has been designed and was said to be due for construction in 1974.

Behind this whole matter of housing and town planning lies the question of how permanent the settlement is expected to be. If the main motivation for Soviet development of the north is mineral extraction, the life expectancy of towns dependent upon it may be short. But there may be other motives: to establish sovereignty, to influence the native peoples, to defend against attack. Until now, permanent settlement has been encouraged, so presumably non-economic motives were seen as important. Now there is a questioning of this view, because permanent towns, with the appropriate social services, have proved expensive, and no doubt because non-economic factors are losing importance. Towns like Noril'sk are particularly vulnerable to this challenge, because an element of prestige building can rather easily be discerned in them. The new proposal is for temporary towns, designed for the expected period of working the ore body, with workers coming up in shifts, without their families, from permanent towns in the south. This view is gaining ground, but is not yet dominant (it would apply only to certain situations, in any case). In fact, it is hotly disputed by some, who call it the 'strategy of double costs' because, they say, in the end towns are built both in the south and in the north. But the so-called shift system is being tried out; while new towns in the southern part of the west Siberian oilfield are being planned on the old scale – Surgut is expected to reach 100,000 in due course, and Nizhnevartovsk 130,000 – the northern part, round the Ob' and Taz estuaries, is to have minimal population and maximum automation. Oilfield workers in the peninsula of Yamal, for instance, were based in 1977 at Tyumen', 1,500 km to the south. If this idea prevails elsewhere, there will of course be radical changes in the housing and town planning program. A view that probably does command fairly general support is that towns, when built, should be larger but fewer; they should serve more than one mine, or industry, and therefore be less vulnerable.

Trade and commerce

Exploitation of the resources mentioned in the earlier sections is undertaken in all cases by the state. The products are marketed under arrangements approved by the state planning commission (Gosplan), with fixed allocations and prices. This centralized control, which applies throughout the country, largely eliminates the play of market forces as manifested in the capitalist world. Large-scale trade and commerce, therefore, become chiefly matters of distribution. Even craftwork by northern natives is in the main centrally controlled, and is produced in such officially established institutions as the industrial combine (*promkombinat*) at Uelen in Chukotka. Casual and relatively uncontrolled trading takes place only in the context of collective-farm workers selling to the public the produce of their own backyards. In temperate areas of the USSR this trading has great importance, but in the north, where the opportunities for small-scale food production are much more limited, its role is small.

Population

The Russians penetrated the northern parts of Europe and Asia during a period extending from the tenth century (the White Sea region) to the seventeenth century (most of northern Siberia). They found native peoples already living in almost all the territory they occupied. The first Russian settlers were hunters, cossacks, and peasants, and they were soon followed by exiles. Their numbers remained small until the twentieth century. The native peoples outnumbered them very easily, but the territory was so vast, and the natives so widely scattered, that it was not hard for the Russians to establish a local superiority in numbers round the settlements. Meanwhile, down the centuries, the number of both Russian settlers and natives grew, and probably remained in much the same proportion overall.

The big change came in the last forty years. The dynamism of the Soviet system led to a vast inflow of settlers from the south. They may be referred to loosely as Russians, but in fact included many Ukrainians and White Russians, besides a smaller number from other Soviet minority peoples, such as Georgians and Tatars. There were also, for a period, a large number of Poles and Germans. The motivation for this inflow was the government's desire for self-sufficiency in the industrialization program, and a notorious feature of it was that a large proportion of the immigrants were under constraint. It is not possible to give any exact idea of the size of these unwilling groups of immigrants. Census results are little help in this respect, because the 1926 census preceded this wave of immigration, and the next one was in 1959, when the worst was already over. (There was a census in 1939, but it was incomplete and was later shown to be inaccurate; probably the difficulty was precisely the prison-camp situation.)

Nevertheless, there was a striking difference between the results of these two full censuses. The total population increased by a factor of between four and five, and, even more significant, the immigrants advanced from being a minority of one to four in relation to the natives to being a majority of four to one. This was due to the arrival of the workers to man the mines, the transport system, the forestry and food industries, the construction industry, the communications network, and, particularly important in a socialist economy, the administration of all this. It is not likely that the figure for the total population in 1959 (see Table 2) included any very large number of persons under constraint; most camps in the north had closed by then. But it is likely that quite a large number of persons included had originally gone north unwillingly, but now stayed on. There could be several reasons for this: the physical difficulty of taking up the old life after being possibly presumed dead for a decade or more, or just the need to earn the north's high wages in order to make up for long years of earning nothing. The truth about all this will not be known until the Soviet government releases the information, and there is no sign at all that it is ready to do so yet.

Since this very large change was revealed in the 1959 census figures, there has been a further census, in 1970, and annual totals are published for the population of major administrative areas. These data show that the trend is continuing, but at a somewhat slower rate. The increase in total population of the north between 1959 and 1976 was about 100,000 a year, or 42 per cent over the seventeen years, which was well over twice the rate of increase in native population of the north. The natives thus became a minority of one to six, or, if the marginally 'northern' and heavily Russian areas in the White Sea drainage are omitted (Karel'skaya ASSR and Arkhangel'skaya Oblast'), one to five.

The incoming Russians are predominantly towndwellers. There were by 1975 nine cities of over 100,000 (see Appendix 1), and at least thirty-nine more between 15,000 and 100,000. They are industrial or administrative centres, and are located in the main regions of Russian immigration. Three-quarters of them are in European USSR. The relatively high population of Murmanskaya Oblast' (Table 2) is attributable to the fishing industry based on Murmansk itself, and to mining south of the town. In Karel'skaya ASSR and Arkhangel'skaya Oblast' forest industries predominate. Komi ASSR has them too, and there is also a major coalfield around Vorkuta in the northeast. The rapidly increasing figure for Khanty-Mansiyskiy NO reflects the growing oil and gas industry there. Noril'sk is a major mining centre. Yakutskaya ASSR has many mines, and Yakutsk

Table 2. Population of administrative districts in northern USSR, 1926–76

Administrative district (1959–75 boundaries; 1926 figures are for roughly equivalent areas)	Population in thousands							
	1926	1959	1970	1972	1973	1974	1975	1976
Murmanskaya Oblast'	23	568	799	835	854	881	905	930
Karel'skaya ASSR	270	651	714	715	720	722	726	735
Arkhangel'skaya Oblast' excl. Nenetskiy NO }	429	1,230	1,363	1,376	1,376	1,382	1,390	1,407
Nenetskiy NO		46	39	39	39	39	40	41
Komi ASSR	207	806	965	984	997	1,012	1,023	1,053
Yamalo-Nenetskiy NO } Khanty-Mansiyskiy NO }	49	62	80	98	103	109	118	126
		124	272	306	332	362	390	425
Taymyrskiy NO }	24	33	38	41	41	42	42	43
Evenkiyskiy NO		10	13	13	13	14	14	14
City of Noril'sk		118	136	145	150	156	161	168
Yakutskaya ASSR }	289	488	664	694	715	736	756	779
Magadanskaya Oblast' excl. Chukotskiy NO }		189	251	269	280	293	304	308
Chukotskiy NO		47	101	112	116	118	122	125
Kamchatskaya Oblast' excl. Koryakskiy NO }	35	193	256	279	289	301	313	323
Koryakskiy NO		28	31	32	33	33	33	34
TOTAL	1,326	4,593	5,722	5,929	6,058	6,200	6,337	6,511

ASSR = Avtonomnaya Sovetskaya Sotsialisticheskaya Respublika (Autonomous Soviet Socialist Republic).
NO = Natsional'nyy Okrug (National District).
The administrative districts listed above, taken together, do not quite constitute the total area of 'the Soviet north', which includes the northern parts of certain other districts. Thus the population of 'the Soviet north' in 1970 was 6.7 million. The difference between the two areas is given in detail in *Polar Record*, Vol. 17, No. 108 (1974), pp. 314–18.
(*Sources: Narodnoye khozyaystvo SSSR*. Moscow, 'Statistika', annually.
Narodnoye khozyaystvo RSFSR. Moscow, 'Statistika', annually.
F. Lorimer, *The Population of the Soviet Union*, Geneva, 1946, pp. 67–70.)

itself is an administrative and communications centre. Magadanskaya Oblast' has more mining, while Kamchatskaya Oblast' is the centre of the Pacific-based fishing industry. All these occupations are carried out almost exclusively by immigrants. It is no doubt immigrants also who mostly man the defence installations which must exist in the north; but their numbers are unlikely to have been included in the regional census returns, so the scale is unknown.

The northern natives, on the other hand, are to be found predominantly in rural areas, following their traditional pursuits of tending animals, hunting, or fishing. But among the three most numerous peoples, which are also the most advanced in the educational and wider cultural sense – the Komi, Yakuts, and Karelians – there are many highly trained and

Table 3. Numerical strength and location of northern peoples in the USSR, 1959 and 1970

Ethnic group (names in current Soviet use followed by other names)	Population in thousands		Per cent change,	Percentage of population who consider language of group to be their native language	
	1959	1970	1959–70	1959	1970
1 Komi (Zyryans)	287	322	+12	89.3	82.7
2 Yakuty (Yakuts)	233	296	+22	97.6	96.3
3 Karely (Karelians)	167	146	−16	71.3	63
4 Nentsy (Samoyeds)	23	29	+26	84.7	83.4
5 Evenki (Tungus)	25	25	—	55.9	51.3
6 Khanty (Ostyaks)	19	21	+11	77	68.9
7 Chukchi	12	14	+16	93.9	82.6
8 Eveny (Lamuts)	9.1	12	+39	81.4	56
9 Mansi (Voguls)	6.45	7.7	+19	59.2	52.4
10 Koryaki	6.3	7.5	+19	90.5	81.1
11 Dolgany	3.9	4.9	+26	93.9	89.8
12 Sel'kupy (Ostyak Samoyeds)	3.8	4.3	+13	50.6	51.1
13 Saami (Lapps)	1.8	1.9	+5	69.9	56.2
14 Itel'meny (Kamchadals)	1.1	1.3	+18	36	35.7
15 Kety (Yenisey Ostyaks)	1	1.2	+20	77.1	74.9
16 Nganasany (Tavgi Samoyeds)	0.75	1	+33	93.4	75.4
17 Yukagiry	0.4	0.6	+50	52.5	46.8
18 Eskimosy (Asiatic Eskimo)	1.1	1.3	+18	84	60
19 Aleuty (Aleuts)	0.4	0.4	—	22.3	21.7
TOTAL	802.1	897.1	+12		
Total Population of USSR	208,827	241,720	+16		

(*Source*: *Narodnoye khozyaystvo SSSR v 1970g*, Moscow, 1971, pp. 16–17.)

qualified persons who can and do fill the jobs otherwise taken by Russians. The numbers of all the northern peoples increased, in the aggregate, by 12 per cent between 1959 and 1970 (see Table 3). The Karelians, however, show a drop of 16 per cent. No explanation has been given for this, but one may guess that the cause was assimilation rather than a disproportionately high death-rate. The Evenki (Tungus) remained static in numbers. The others all increased by amounts varying from 5 to 50 per cent; but the percentage increase for the peoples with low absolute numbers may be rather inaccurate, because the population statistics released are given only to two significant figures.

There is an unexplained contrast here with the North American situation. In Canada, the Eskimo population has increased in the same period by 40 per cent, and the situation seems to be closely similar in Alaska and Greenland. The reason is improved medical services and a very high

Table 3. Numerical strength and location of northern peoples in the USSR, 1959 and 1970

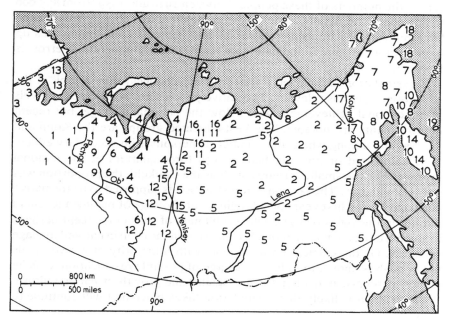

Distribution of northern peoples in the USSR, 1970. The numbers refer to the left-hand column in Table 3 opposite.

(*Source*: *Atlas SSSR*, Moscow, 1969, pp. 98–9.)

birth-rate. The absolute sizes of those three Eskimo populations are in the range 10,000–40,000. Taking the Soviet northern peoples in this range (Nentsy, Evenki, Khanty, Chukchi and Eveny), the increase is only 15 per cent, or less than half. If all the other smaller Soviet northern peoples are taken into account too, the percentage increase drops to 14 per cent. It is hard to imagine that medical services analogous to those in the West are not also available. The number of doctors per head of population in the Soviet north is stated to be considerably above the figure for the whole country. Perhaps the explanation is to be found in a higher assimilation rate. The 1970 census does show that there is a steady drop in the proportion of each people which regards the language of that people as its native language. In only one case was this not evident – the Sel'kupy, where the proportion rose marginally. In some of the others – the Eveny and the Nganasany – there was a drop of around 20 per cent. But this evidence is inconclusive, and the full answer can come only from Soviet investigators.

An even odder situation is apparent from study of the figures from the 1926 census, which are also available and reveal an overall drop of 4 per cent in the numbers of these peoples between 1926 and 1959 (from 835,000 to 802,000). The details have not been included in Table 3, because the large number of astonishing increases and decreases, in some cases up to a factor of three, indicate that substantial enumeration error is possible.

If assimilation has played a part in these changes, then a look at the education system is relevant. The native peoples are in all cases at levels of educational achievement well below the national average, but recent improvements are notable. The best-educated groups are the Komi and the Yakuts, among whom the number of persons with secondary education, completed or not, reached in 1970 three-quarters of the national average. For the smaller groups, such as the Chukchi, the proportion was only one-quarter. The language of instruction is supposed to be the native language for the first years of schooling, but shortage of qualified teachers did not in the 1960s or 1970s allow fulfilment of this aim, except perhaps in some Komi or Yakut schools. We have noted the drop in native-language speakers. Of the curriculum in schools with a majority of native pupils not enough is known to judge the degree of possible russification. Official policy, it seems, is neither for nor against the growth of native culture. If this is so, it is likely that assimilation takes place and will continue to do so.

Labour

Fundamental to any plan for economic development of the Soviet north is the provision of a labour force. When the Soviet government first decided to undertake large-scale development, the region's main inhabitants were the thinly distributed native peoples, who were not only too few in number at any given point, but lacked both the necessary skills, and perhaps the inclination, to work in the new enterprises. Russians from outside had to be brought in. These were either volunteers, attracted to work in the north for a variety of reasons, or forced labour, sent to the north both as a punishment and in order to fill the gap. Each of these three groups will be discussed.

The forced labour may be dealt with first, since its use in the north for economic purposes was, one may hope, transitory. The period during which it flourished in the north was roughly the thirty years between 1929 and 1958. There are many categories of forced labourers: inmates of corrective labour camps or of corrective labour colonies, 'special settlers' (who often included released prisoners from the first two categories), exiles, prisoners of war. Jointly, they constituted a large part of the labour

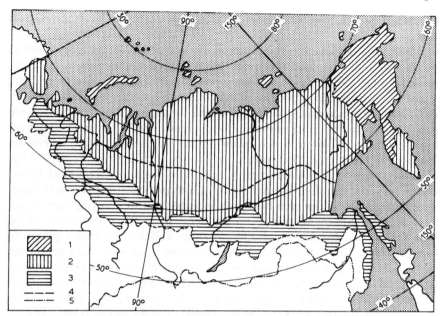

Map 8. Regional wage differentials in the Soviet north, 1960 and 1968. Zones in which regional wage differentials were provided under the decree of the Supreme Soviet of the USSR of 26 September 1967, effective from 1 January 1968
Key: 1, 'the far north': increment of 10 per cent every six months to ten increments (100 per cent maximum). 2, 'the far north': increment of 10 per cent every six months to six increments, then 10 per cent every year to two increments (80 per cent maximum). 3, 'localities equated to regions of the far north': increment of 10 per cent every year to five increments (50 per cent maximum). 4, boundaries of zones under previous decree of 10 February 1960, by which rates of increase were also slower. 5, international boundary.

force for the most important mining developments: the Vorkuta coalfield, the Noril'sk nickel mines, the Aldan and Kolyma goldmines. The Kolyma region was run by Dal'stroy, an organization openly controlled by the NKVD and relying on forced labour. The number of persons under constraint altogether, or at any given time, is not known with any accuracy, but is certainly of the order of hundreds of thousands. Very large numbers died. As a labour force, this one was of course extremely inefficient. It appeared to be cheap, but must in reality have been expensive due to costs of guarding. Nevertheless, it succeeded in starting (at fearful human cost) a series of enterprises which still continue, and which probably would not have started at all in its absence. The death of Stalin in 1953 was the turning point in this system, and numbers dropped very considerably thereafter. It is not likely that the north is now entirely free of this form of labour, but its large-scale economic use does seem to have come to an end.

When this happened, an alternative method of attracting labour had to be found, and this was the more usual one of higher wages. In fact, regional wage differentials had been paid since 1932, because there were always certain kinds of work for which forced labour was not suitable. But they came into much wider use in the 1950s, and their rate and scope has been adjusted at intervals (1945, 1960, 1968) to meet changing circumstances. The main provision of the statute has always been an increase in basic pay. The current rates vary with the remoteness of the area (see Map 8). For the remotest zone, there is an increment of 10 per cent every six months to a maximum of 100 per cent; for the next zone, an increment of 10 per cent every six months for three years, then 10 per cent a year, to a maximum of 80 per cent. Both these zones are called 'the far north'. For the third, closest zone, called 'localities equated to regions of the far north', the increment is 10 per cent a year to a maximum of 50 per cent. The sliding scale is, of course, designed to retain as well as attract labour, and it constitutes one of the most interesting features of the legislation. Other benefits include an initial cash grant and free travel to the north for the worker and his family, earlier pension entitlement, longer annual leave with some free travel, and the right to retain accommodation in the south. Since 1960, all these provisions apply not only to immigrant labour, but to persons already resident in the north, including natives. Quite apart from these incentives, there are cost-of-living allowances, known as regional coefficients, which also vary from place to place.

The third group is the native peoples. Obviously, they can be regarded as a potential labour force, which would be cheaper than any other because resident relatively close to the place of work. But until recently there has been little evidence that any large number of natives went into industrial work. The great majority continued to follow their traditional pursuits of hunting, fishing, or tending reindeer or other domesticated animals. It has in fact been official policy to encourage this, since these were jobs which incoming southerners would have greatest difficulty in doing. But recently it has become clear that the standard of living of the pastoral group has been lower than that of the industrial workers, so there is a feeling that a freer choice should be offered to natives. The latter became eligible for the northern increments and associated benefits in 1960, but, while state-farm workers received them, collective farmers did not, since they are not classified as wage-earners.

Despite the incentives, the turnover of labour has continued to be higher than it should be for full efficiency, and there is a shortage of labour in some regions. The proportion of industrial workers who left their jobs each year in Yakutskaya ASSR between 1955 and 1962 varied between a third and a half. At particular mines in remote districts the proportion approached 100 per cent. In the years 1958–68 in the Soviet northeast, 80 per cent of the population left and was replaced by new immigrants. The severe climate and the remoteness and isolation were no doubt major factors in causing high turnover. But there were other factors too, over which some control is possible. Chief among these was the low standard of living and the shortage of many amenities. There was official awareness of this, as evidenced by Khrushchev's remark about developing regions in 1959: 'New people must be attracted not by the high pay by comparison with other towns, but by good living conditions.' Unfortunately, this thinking led only to a reduction of the northern increments in 1960, and not to any very noticeable improvement in living conditions. The increments had therefore to be increased again in 1968. The long-term intention is no doubt to improve living conditions, but this is inevitably a slow process. It is worth noting that the annual turnover of labour in Noril'sk is only 13 per cent, which is the average for the RSFSR as a whole. Magadan, another new town, may be comparable in this respect. But in most other places, it seems that there is still a long way to go before conditions approach those in the south. Meanwhile, standards in the south are improving, so that still further effort is required in the north to catch up.

Of the social life which grows from this background little is accurately known, for there are few reliable published accounts. It seems clear, however, that there is a fairly close parallel with northern settlements elsewhere: a high turnover rate and a disproportionately large number of young unmarried males among the immigrant population lead to a basic instability. This produces such problems as alcoholism and a divorce rate twice that in the south. No crime statistics are published in the USSR, but there is no reason to think the crime rate may be lower in the Soviet north than in other parts of the north. The chief motive for going north in the USSR is – in about half of all cases – the extra money. Studies show that about 15 per cent of immigrants to the north have higher-minded motives, and these persons tend to stay five times as long as the others. On the subject of Russian–native relations, even less is known. The guess here would be that they are generally good, especially now that there is equal pay for equal work. A certain condescension by Russians towards natives has been observed, but this is usually harmless enough.

Government and administration

The northern parts of the USSR are administered in just the same way as are other parts of the country. The breakdown into administrative subdivisions follows the same principles there as elsewhere. This has been broadly true throughout the Soviet period, although there was a somewhat different arrangement in the 1930s, of which more will be said below. It was also true under the Tsars.

The administrative system at present in force has been unchanged in principle since the early years of the Soviet period. The multinational character of the state caused provision to be made for two sorts of administrative unit: those based on national criteria, to provide political recognition and some cultural autonomy for minority ethnic groups; and those without any such basis. The first-order divisions within the USSR are the fifteen union republics, which are national-based. The whole of the northlands falls within the largest of these republics, the Russian (RSFSR). Below that level there are on the one hand autonomous republics (*avtonomnaya sovetskaya sotsialisticheskaya respublika*, abbreviated to ASSR), autonomous provinces (*avtonomnaya oblast'*, abbreviated to AO), and national districts (*natsional'nyy okrug*, abbreviated to NO), three national-based units in descending order of importance; and on the other, territories and provinces (*kray* and *oblast'*), which are of broadly equal standing with each other and are at the same level as autonomous republics. Below all these are the regions (*rayon*), roughly equivalent to counties in Britain and townships in North America, and common to both national and non-national ladders.

All these territorial units are run by elected soviets. Those of autonomous republics are rather more elaborate and are called Supreme Soviets and Councils of Ministers, as in the union republics. The national-based units send deputies to the Soviet of Nationalities, a chamber of the Supreme Soviet of the USSR, in Moscow: a union republic sends twenty-five, an autonomous republic eleven, an autonomous province five, and a national district one. The other chamber, known as the Soviet of the Union, is filled by deputies elected on a nation-wide basis from electoral districts of about 300,000 inhabitants each.

These organs exist and function. But real power resides not so much in them as in a parallel but until 1977 non-constitutional hierarchy of committees of the Communist Party of the Soviet Union (KPSS). Each territorial soviet has an equivalent party committee, which frequently has overlapping membership and may even occupy the same building. The chairman of the territorial soviet is almost always a member (but not the chairman) of the corresponding party committee.

The north of the RSFSR comprises the whole or parts of fifteen second-order divisions (see Map 9). Three are national-based: from west to east, Karel'skaya ASSR, Komi ASSR, and Yakutskaya ASSR. The others are Murmanskaya Oblast', Arkhangel'skaya Oblast', Tyumenskaya Oblast', Tomskaya Oblast', Krasnoyarskiy Kray, Irkutskaya

Oblast', Chitinskaya Oblast', Amurskaya Oblast', Khabarovskiy Kray, Magadanskaya Oblast', Kamchatskaya Oblast', and Sakhalinskaya Oblast'. Below these again are third-order divisions, which include seven national districts: Nenetskiy NO within Arkhangel'skaya Oblast', Yamalo-Nenetskiy NO and Khanty-Mansiyskiy NO within Tyumenskaya Oblast', Taymyrskiy or Dolgano-Nenetskiy NO and Evenkiyskiy NO within Krasnoyarskiy Kray, Chukotskiy NO within Magadanskaya Oblast', and Koryakskiy NO within Kamchatskaya Oblast'.

National-based units of the second and third order (ASSR and NO) occupy, it may be noted, about four-fifths of the territory north of latitude 60° N – a much higher proportion than in any other part of the USSR. This arrangement has been made in recognition of the fact that there are a number of different minority peoples in the north. At the time of the creation of these units (between 1921 and 1930), the minority peoples for whom they are named formed a significant proportion of the total population, in some cases a clear majority. But in the subsequent years there has been a massive inflow of immigrants from the south, mainly of Great Russian origin, and the northern peoples are now everywhere in the minority in their own regions, often a rather small minority. Their voice in the management of the affairs of the national-based units is, however, still strong. It is appropriate to add here a reminder that the word 'autonomous' used in the title of the larger national-based units does not imply any independence from Moscow. Rather, it should be taken to mean that the people concerned are granted by the constitution certain rights in the ordering of their own affairs. These rights are by no means an empty formula; but they cannot be interpreted, by those who enjoy them, as a licence to adopt policies not approved by the Soviet government. That would be nationalist deviation, and those who attempted it in the early days of the Soviet regime were harshly repressed. Moscow retains its control by two chief means: selection of administrators, through the cadres branch of the Communist Party, and central direction of the major industries.

It will be noticed that the territorial units at *kray*, *oblast'*, and ASSR level are not so organized as to conform neatly with any of several possible southern boundaries of 'the north'. Some are wholly within it, some straddle the boundary. Much less is there any single 'northern' unit. However, there have been administrative structures of an economic kind superimposed on the territorial units. In the 1930s the largest structures of this kind were the Chief Administration of the Northern Sea Route (abbreviated to Glavsevmorput'), and the Chief Administration of Construction in the Far North (abbreviated to Dal'stroy). The former was set up in 1932 as a shipping agency, with the task of developing the northeast passage as a commercial waterway, but for a few years (1935–8) it had

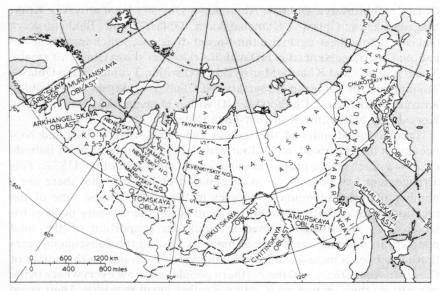

Map 9. Administrative divisions in the Soviet north, 1976. The boundaries have remained stable for several decades, with virtually no change since 1953

responsibility also for many economic and social services on land. Its territory then was enormous, extending southwards to latitude 62° N in much of Siberia. The latter, Dal'stroy, was in existence from 1931 to 1953, but was more confined in area. It was a mining organization active in the Kolyma basin, with rights extending into many fields other than mining, but it was never much publicized, because it was run by the Ministry of Internal Affairs and relied almost exclusively on forced labour. This made it not only mysterious but notorious.

In 1957 there was a major reorganization of the country's economic management apparatus. With the object of decentralizing, a series of economic administrative regions were set up, each managed by a local economic council (*sovnarkhoz*). These regions were agglomerations of existing administrative regions; some were wholly in the north, but most straddled the southern boundary. An interesting development was the creation of a wholly northern *sovnarkhoz* in the far northeast, consisting of Yakutskaya ASSR and Magadanskaya Oblast'. But in 1965 the *sovnarkhoz* system was swept away in a new reform, which effectively restored the previously existing arrangement, but with some modifications.

Yet another regional grouping which started to emerge in the 1950s was that of major economic regions. These have no executive structure or function, but are used by Gosplan for planning and statistical purposes.

The north falls wholly into four of these – the northwestern, which comprises the European north and the Leningrad region, the West Siberian, the East Siberian, and the Far Eastern, which includes the whole Pacific coast (Yakutskaya ASSR was at first in the East Siberian region, then from 1966 in the Far Eastern). But all four include territory which does not, by any definition, form part of 'the north', and indeed the most populous and developed parts of each are at the southern end.

In summary, there is no pattern of administrative subdivisions in the USSR which reflects any attempt to bring the whole of the Soviet north, or even major parts of it, into a single unit. The nearest this came to happening was in the case of Glavsevmorput'. Nevertheless, there is quite an influential body of opinion which favours the idea of moving in the direction of a unifying administrative authority for the north. The Inter-departmental Commission on Problems of the North, an advisory body set up in 1963, has had such views expressed to it. The usefulness of such an authority, in co-ordination, in avoidance of duplication, and in introduction of standardized procedures appropriate to the northern environment, is quite clear. Yet it is also clear that major communications links will continue to run north–south, as they do in Canada; Yakutsk is likely to remain more closely linked to Irkutsk than to, say, Noril'sk, which is a comparable distance away. Indeed, one reason for the north–south orientation was in order to permit the economic potential of the better-developed south to help develop the north. Further, any new authority would have the delicate and difficult task of displacing firmly entrenched existing authorities. So it would seem that any future unifying authority for the north is likely to have mainly a co-ordinating function.

Transport

The transport system is the key to the development of any northern territory, where sparseness of population and great distances imply absence of any pre-existing transport network. The relative importance of the various transport media within the system is also different in the north, in that land transport tends to drop from first to last place. In the Soviet north, the order is water, air, land.

Water

The waterways were for long the only highways of Siberia. The first Russians to enter Siberia used them. Their importance remains great today, because of their ability to carry heavy freight, and because there happens to be a good network of major rivers. Two features of this network have special importance. Most of the rivers flow northwards, and thus provide a way into the northlands either from the road and rail arteries of the south, or from the Arctic Ocean; and all of them freeze over for a substantial part of each year. The seasonality can be turned to some advantage, since the ice can be made to provide a highway for vehicles.

The major arteries today are the Ob'-Irtysh, Yenisey, and Lena, all of which have connection with the Trans-Siberian railway. The Ob' and Lena have relatively shallow mouths, and thus exclude ocean-going shipping, but the Yenisey will admit ships of up to 10,000 tonnes displacement as far as the timber port of Igarka, 673 km upstream. All are navigable for almost the whole of their length. The Severnaya Dvina and Pechora, in European USSR, are lesser, but still large, rivers which are also intersected by the railway. Apart from these, there are several large navigable rivers on which tugs and barges operate: the Taz, Khatanga-Kheta, Olenek, Yana, Indigirka, and Kolyma. The shipping season varies from six months in the south to three at the mouths of these rivers. Once the ice has gone out in the spring, it will not return (unlike the situation at sea) until autumn. But ice is not the only hazard. Although the incidence of rapids is low, water levels can drop by a very large fraction in the summer and autumn, complicating navigation. The quantity of traffic and of freight carried cannot be determined precisely, since the southern boundary of 'the north' does not generally coincide with a unit for which figures are quoted. But it is of the order of tens of millions of tonnes a year. The waterways of the Lena basin, for instance, carried 5,479,000 tonnes in 1970. It is likely that the Yenisey below Yeniseysk carried a comparable amount; and the Ob'-Irtysh, within the same latitude bracket, probably carried more, in view of the oil and gas development (the port of Surgut alone was due to handle 2 million tonnes in 1973). The total for the whole country in 1970 was 358 million tonnes.

The equipment used includes barges of up to 2,150 tonnes capacity, either self-powered, or towed or pushed by tugs. Water depth is the major problem. The Yenisey below Krasnoyarsk has a guaranteed depth of only 2.4 m, and the other rivers have less. Passenger vessels include standard types, found in large numbers on the rivers of European USSR, but in particular hydrofoils, predominantly the 'Raketa' model. There are repair yards and some building facilities on the bigger Siberian rivers, but many of the larger ships assigned to them are built in European USSR, or Eastern Europe, and make the sea voyage via the Arctic Ocean to their destinations.

The mouths of all these rivers are linked by the waters of the Arctic Ocean – the old northeast passage, renamed in the USSR the northern sea route. It was this linkage which led the Soviet government to expend much money and effort on making the seaway here navigable. The idea of the through route, connecting Atlantic and Pacific, and greatly reducing the length of a voyage from Murmansk to Vladivostok, was also in mind; but getting at the natural resources of Siberia by way of the rivers was the more important motive. Though the route had been explored and sailed through long before Soviet times, by both Russians and foreigners, it was only after 1917 that technology advanced to a stage at which the overcoming of the ice and other natural obstacles became a practical possibility. Glavsevmorput' in the 1930s had under its control a flotilla of icebreakers, many merchant ships, some of them strengthened for ice navigation, a squadron of aircraft, a chain of weather stations, a hydrographic service, and a major scientific institute (the Arctic Institute at Leningrad). It also had, in its early years, wide control over all Siberia north of latitude 62° N, and at that time it employed 40,000 persons.

Having mobilized resources on this scale, the Soviet government was able to achieve considerable successes. Ships plied between the western termini at Murmansk and Arkhangel'sk and the Ob' and Yenisey rivers, between the eastern terminus at Vladivostok and the Kolyma river, between both and the Lena river. Ports were established near the mouths of these rivers, so that trans-shipment could be effected. There were setbacks. The worst was in 1937, when twenty-six ships were forced to winter at sea, including seven of the eight then serviceable icebreakers. Nevertheless, much was learnt, and between 100,000 and 300,000 tonnes of freight were lifted each season from 1933 to 1938. During the war, activity continued. Murmansk and Arkhangel'sk were, of course, vital bases in a wider context, and the northern sea route accounted for only a small fraction of their activity. Nevertheless, traffic in the Kara Sea continued at a level which caused the German navy to mount U-boat attacks and, in 1942, to send *Admiral Scheer* on a partially successful raid. The eastern end of the route became a useful point of access for lease-lend ships operating out of North American west-coast ports. The result of all this was that the tonnage carried increased during the war.

The position today is that there has been further marked increase. There are no official figures for freight turnover, but it seems likely that this is now of the order of one and a half to three million tonnes a year, carried in 200 to 400 freighters, many of which make two or more voyages a season. The pattern is substantially the same. The traffic from the west to and from the Yenisey river is generally the largest single item. Stores are taken to Dudinka for the mining town of Noril'sk, ore leaves Noril'sk for the west, and timber is taken out from Igarka. The Lena is still served from both east and west, but with the west predominating. The Kolyma now has many mines in its basin, and some tens of ships are involved annually in their resupply. The coast to the east, with the port of Pevek, has also become an important mining region. For a period in the middle

1960s the Ob' attracted sea-going traffic in connection with oil and gas exploration in its basin. It is noteworthy that there is not generally any significant through traffic from one end of the route to the other.

While this pattern of freight movement has continued for a number of years, it is also true that the service area of the northern sea route has steadily decreased in size. Improvements in the transport service from the south, by rail and river, have reduced the area to which the northern sea route can provide a more economic service to a relatively narrow strip along the north coast. The reduction in size of the service area has probably been offset by the growth in volume of freight generated by places remaining within it, so the total turnover is likely to be rather static, or perhaps slowly growing. On the organizational side, Glavsevmorput' was placed under the Ministry of the Merchant Fleet in 1953, and then lost more of its old identity, including its name, in later internal reshuffles. But its operations continue as before, as no doubt do many of its staff. In 1971 its name was partly revived with the creation within the Ministry of a new Administration of the Northern Sea Route (but this one an *administratsiya*, lower in rank than the earlier *glavnoye upravleniye*).

These operations are made possible by a fleet of about fourteen ice-breakers of over 10,000 shp, eleven of them Finnish-built, augmented by up to fourteen smaller icebreakers of about 5,000 shp (Table 4). The atomic-powered *Lenin* was for long the most powerful (44,000 shp), and is Soviet-built. She entered service in 1960, and was the first example of nuclear propulsion in any surface ship. But she was out of action from 1967 to 1970, probably with reactor trouble. This seems to have caused delay in the plans for building more ships of this kind, for, despite announcements of intent as early as 1964, another atomic-powered ice-breaker came into service only in 1975. This was the *Arktika* (75,000 shp), which was not only more powerful than *Lenin* but also incorporated many changes in design. The Finnish shipyard of Wärtsilä built three ships in the *Kapitan*-class of 10,500 shp (1954–6), five ships in the *Moskva*-class of 22,000 shp (1960–9), and three ships in the *Yermak*-class of 36,000 shp (1974–6).

A third atomic-powered icebreaker, *Sibir'*, a sister ship to *Arktika*, is being built in the USSR and is expected to enter service in 1977. These considerable reinforcements and plans indicate much confidence in the future of the northern sea route. The *Arktika*, in a dramatic display of her power, made a successful voyage to the North Pole in August 1977 – the first surface ship ever to do so.

The Soviet freighter fleet available for use in the Arctic is also large. Navigation in Arctic waters is permitted for ships in the top two classes (out of six) in the Soviet ice classification for shipping, and these two classes contained in 1971 over 800 ships. Probably 350 of these have been

Table 4. Soviet icebreakers used on the northern sea route,[a] in service or on order, 1976

Ship	Where built	When completed	Power shp	Displacement tonnes
Arktika	USSR	1975	75,000	23,400
Lenin	USSR	1959	44,000	16,000
Yermak	Finland	1974	36,000	20,240
Admiral Makarov	Finland	1975	36,000	20,240
Krasin	Finland	1976	36,000	20,240
Moskva[b]	Finland	1960	22,000	13,290
Leningrad	Finland	1961	22,000	13,290
Kiyev	Finland	1965	22,000	13,290
Murmansk	Finland	1968	22,000	13,290
Vladivostok	Finland	1969	22,000	13,290
Kapitan Belousov	Finland	1954	10,500	5,360
Kapitan Voronin	Finland	1955	10,500	5,360
Kapitan Melekhov	Finland	1956	10,500	5,360
Sibir'[c] (ex Iosif Stalin)	USSR	1938	10,000	9,300
Vasiliy Pronchishchev (ex Ledokol-1)	USSR	1961	5,400	2,500
Afanasiy Nikitin (ex Ledokol-2)	USSR	1962	5,400	2,500
Khariton Laptev (ex Ledokol-3)	USSR	Before 1963	5,400	2,500
Vasiliy Poyarkov (ex Ledokol-4)	USSR	1963	5,400	2,500
Yerofey Khabarov (ex Ledokol-5)	USSR	1963	5,400	2,500
Ivan Kruzenshtern (ex Ledokol-6)	USSR	1964	5,400	2,500
Vladimir Rusanov (ex Ledokol-7)	USSR	1964	5,400	2,500
Semen Chelyuskin (ex Ledokol-8)	USSR	1964	5,400	2,500
Yuriy Lisyanskiy (ex Ledokol-9)	USSR	1965	5,400	2,500
Petr Pakhtusov	USSR	1966	under 5,000	
Georgiy Sedov	USSR	1967	under 5,000	
Fedor Litke	USSR	1970	5,400?	
Semen Dezhnev	USSR	1972 or 1973	5,400?	
Ivan Moskvitin	USSR	1972 or 1973	5,400?	
On Order, 1976				
Sibir'	USSR	1977	75,000	23,400
M. Ya. Sorokin	Finland	1977–8	20,000 ⎫	shallow draught
N. M. Nikolayev	Finland	1977–8	20,000 ⎬	vessels
Otto Schmidt	USSR	1978	5,400	3,650

[a] These are ships known to have been used in the Arctic. The USSR has other icebreakers, particularly in the 5,400 hp class.

[b] One source (*Morskoy Flot*, No. 11, 1975, pp. 42–5) gives rather higher tonnage and hp figures than do most others: *Moskva* class – 15,000 tonnes and 26,000 hp; *Yermak* class – 21,000 tonnes and 41,400 hp; *Lenin* 19,500 tonnes; *Sorokin* and *Nikolayev* – 24,800 hp. The hp figures are evidently engine capacity rather than shaft hp.

[c] *Sibir'* (ex *Iosif Stalin*) will no doubt be withdrawn from service before the atomic-powered *Sibir'* becomes operational.

(*Source: Polar Record* and *Inter-Nord*, *passim*, based on Soviet press reports.)

used at some time on the northern sea route. They are of small cargo capacity, most being between 3,000 and 9,500 dwt, but include two container ships of 13,000 dwt (about the largest ships known to have used the route). No doubt shallow water imposes these restrictions.

The shipping season throughout the 1950s and 1960s lasted about four and a half months (early July and mid-November) in certain sectors, but less than three months for the through route. Efforts were then made to extend the season at each end. Every year since 1970 late voyages have been made from Murmansk to the Yenisey and back, leaving in November or December and returning to Murmansk in December or January. The latest of all was that of December 1972, when the ships returned in late January 1973. Conditions were very difficult, with near-total darkness and ice rendered more resistant by snowfall, so that crossing the Kara Sea took much longer than usual, even with five icebreakers in attendance. After this experience the winter voyages ended around New Year's Day. Such voyages seem to have become part of the regular pattern, so the season in that sector is effectively lengthened to six months. Efforts to open the season earlier started with spring voyages to the Kara Sea in 1976. Ships under icebreaker escort successfully off-loaded freight on to the fast ice off the west coast of Yamal, where shallow water prevents ships from coming close inshore. Such voyages were made in April 1976 and in March and April 1977. Operation was difficult, but the job was done. Whether these voyages imply that ships could go elsewhere in the Kara Sea and beyond at this time of year is not yet clear.

The question arises as to whether year-round operations may now be in sight. Powerful reinforcement of the icebreaker fleet has been received, with more on order. Its possible impact is hard to judge. There has been talk of further extension of the season, notably by writing an intention to that effect into the tenth five-year-plan (1976–80), and there may be the possibility of making advances, particularly in the Kara Sea. But perhaps the most promising development in this direction was the voyages of the US supertanker *Manhattan* in Canadian Arctic waters in 1969 and 1970, which demonstrated that really big ships can overcome ice more effectively than had been supposed. Still bigger ships, designed for ice and with a displacement ten or even twenty times greater than today's icebreakers, would have great potentiality, combining tremendous icebreaking power with large freight capacity. However, limitations of water depth in much of the sea north of the USSR effectively precludes operation of such ships along the northern sea route. In this context the construction of two shallow-draught icebreakers of 20,000 shp might be important. Soviet press comment has drawn attention to them, and has expressed hopes that they could help usher in year-round operation. That seems unlikely, but they may help marginally. All these attempts to lengthen the season should be seen against a background of steadily worsening ice conditions. A cooling period which started in the 1940s and is particularly evident in the Kara Sea region is now expected by most Soviet sea-ice specialists to last into the 1990s.

The route has always been a facility of interest almost exclusively to Russians. It can be regarded, however, as a waterway of potential international significance. Although Soviet jurists have sought to show that its component seas were 'internal waters', and although right of innocent passage through territorial waters was on one occasion refused to a US icebreaker (in 1967), the Soviet government has in fact never denied the route's status as a seaway open to all nations. But only limited portions of the route have been used by non-Soviet ships. In early 1967 the Soviet government offered, for a fee, to open the through route to foreign shipping by providing the necessary icebreaker and navigational services, and pointed out that its use could save up to thirteen days' steaming time between, say, Hamburg and Yokohama. But none accepted the offer, since it was clear that factors such as fog or icebreaker unavailability might seriously lengthen the voyage. After the Middle East war of June 1967, it is believed that the offer was tacitly withdrawn, for political reasons. In the longer-term future, it would seem that if a demand develops among the world's shipping interests for a shorter route between North Atlantic and North Pacific, that demand is more likely to be satisfied by a direct route across the Arctic Ocean, using as yet unbuilt very large ships, than by the northern sea route.

Air

In the space of the last generation, air transport has become the standard medium for conveyance of persons and light freight. The flight network in the north grew directly out of that for the rest of the country. The environment itself caused few special problems. There is no aerial counterpart to floating ice, and air temperatures lower than those at ground level in the Arctic are found at high altitude in temperate or tropical regions, so the need for special equipment was much reduced.

In the early days the passenger and freight routes to and in the north were operated by an organization called Polar Aviation, which was a subdivision of Glavsevmorput'. In 1960, however, the state airline Aeroflot took over responsibility for Polar Aviation's activities, and the unit was absorbed into Aeroflot.

Unfortunately, almost no statistics are published about Aeroflot's operations. As the only airline in the USSR, and one which also operates on many international routes, it must surely be the biggest in the world in terms of passenger/km flown. Timetables and route maps are published only for the more important routes, on which foreigners may travel. From these it is clear that there is a frequent service between Moscow and all large cities. This service embraces in the north such centres as Murmansk, Arkhangel'sk, Syktyvkar, Noril'sk, Yakutsk, Magadan, and Tiksi. But below this level of long-distance connections, there is a much denser network of local flights, and it is these which hardly appear at all in published lists. Such information as is available does indicate a northern network which includes all inhabited places down to village level, at a frequency between several flights a day and one a week. The feeder-lines from the smallest places are serviced by single-engined AN-2 or YAK-12 aircraft, flying to regional centres. From here twin-engined IL-14, LI-2, and AN-24 and the tri-jet Yak-40 make the longer flights to provincial capitals (for instance, Yakutsk), whence the big jets and turbo-props – TU-104, IL-18, AN-10 – fly across the country. Soviet passenger aircraft seem to operate with a high load factor, explained partly by the relative cheapness of flying, and perhaps partly by a tendency to cancel flights if they are too lightly booked. Aeroflot also makes available aircraft (including helicopters) and crews on charter, as well as operating a flying medical service. Aeroflot has links with, but is quite distinct from, the Red Air Force, which operates in the north but has little contact with other civilian organizations.

It is not possible to determine a figure for the numbers of passengers carried on northern routes. As a rough guide, however, the following may be helpful: in about 1970 four northern regional directorates of Aeroflot – those based at Syktyvkar, Tyumen', Yakutsk, and Magadan, serving over half the area of the Soviet north – are reported to have carried some 4.5 million passengers, out of a national total for that year of 83 million.

For freight services, the USSR employs the AN-10 (originally a cargo aircraft, but convertible to passenger use), and the massive AN-22, whose payload is four times that of the C-130 Hercules. Much use is also made of the MI-6 helicopter, for long the biggest in service anywhere.

Hovercraft have been coming off the production line in the USSR since 1969. Their potentialities for the north are appreciated, and tests have been conducted there with the 'Gorkovchanin' model. They are not yet in wide use in the north, however, possibly because of the easier availability of helicopters and of hydrofoils.

There can be no doubt that an efficient air service is provided for all parts of the Soviet north. Many places, as in northern Canada and Alaska, can be reached only by air. One of the most striking developments made possible by air transport has been the opening-up of the oil- and gasfields in northwest Siberia, a region with no road or rail connection from the start of development in 1953 to 1970, and with only seasonal water links to such of its centres as were close to rivers. Work in the oilfields themselves was made possible by using very large numbers of helicopters.

Land

Overland transport routes, in the form of tracks through the forest, have existed for as long as human beings have lived in the north. But roads able to carry vehicles, and railways, are relatively recent creations. The network of neither of these is yet very extensive, since the high cost of both construction and maintenance cannot be easily justified in so sparsely inhabited a region. The stimulus to build them has come in almost every case from mineral development.

All-weather roads are few. There are several in the western part of the Soviet north – to Murmansk (completed only in 1968), Arkhangel'sk, Mezen', and parts of Komi ASSR. But east of the Urals there are only two major road systems. One, completed in 1970, runs from Tyumen' to Nizhnevartovsk, an oil centre on the middle Ob'. The other, completed in 1962–4, but with some sections dating from 1942, runs from Bol'shoy Never on the Trans-Siberian railway to the Lena opposite Yakutsk, and on through the mountains to Magadan, with spurs to mining areas. This last, known as the Magadan highway, has a dirt surface, permitting reasonable but not high speeds on most stretches, but the two major rivers on its route – the Aldan and the Lena – are unbridged. The central stretch, between the Lena and the Aldan at Khandyga, where the mountains start, is a winter road only. Thus there is little freighting from the railway to the Magadan region. There are other, relatively short, stretches of all-weather road, which do not join up with the national network: for instance, from Zelenyy Mys, a port on the lower Kolyma, to Bilibino, a gold-mining centre, and from Egvekinot, a port on the Bering Sea, to Iul'tin, where tin is mined. The sparseness of roads stems not only from the great distances, but from a Soviet policy decision taken in the early days of the regime to favour other media in preference to motor traffic.

Wide use is made of winter roads (tracks with no surfacing, and therefore usable only in snowy or dry conditions). Most follow well-established routes, and therefore the trail does not have to be cleared very much each year. It is possible to operate traffic on the ice of the frozen rivers, but this is not often done. The winter road network in eastern Yakutskaya ASSR complements the permanent road system. A spur brought into use between 1965 and 1971 carried the system to the shore of the Arctic Ocean at Nizhneyansk.

There is some use of cross-country vehicles. Both tracked models and models powered by airscrews have been produced in the USSR, but it is not clear to what extent they play an economic role. Very large cross-country vehicles with high ground clearance, a type developed somewhat in Alaska, have not been introduced.

Railways have always been preferred to roads by Soviet planners. There are two major lines to the north: to the port of Murmansk, completed in the First World War and extended in 1960 to Pechenga (formerly the Finnish Petsamo); and to the coal mines at Vorkuta in Komi ASSR, completed in the Second World War, and extended in 1958 to the mouth of the Ob' river at Labytnangi. A spur from the Vorkuta line, from Sosnogorsk near Ukhta to Troitsko-Pechorsk, primarily for timber, came into use in 1975; it is envisaged as the first stage of a line to Solikamsk, the northern railhead in the western Urals. Another, shorter, spur, from Synya to Usinsk (Ust'-Usa), serving the oil industry, was under construction in 1976, and there was talk of its extension later to Nar'yan Mar at the mouth of the Pechora. East of the Urals, however, there is no major line to the north. But there are a number of spurs striking northwards from the Trans-Siberian line, generally planned with the object of opening up new timber-felling areas. From west to east, these are Ivdel' to Sergino on the Ob', Tavda to Leushi on the Konda, Tyumen' to Surgut, Asino to Belyy Yar, Achinsk to Lesosibirsk (formerly Maklakovo), Reshoty to Boguchany, Khrebtovaya to Ust'-Ilimsk. Most of these are 200–300 km long, and most were completed in the 1960s. The only one still uncompleted in 1975 was Reshoty-Boguchany, delayed by changes in the location of the dam it was to serve.

The line from Tyumen' to Surgut, which is 700 km long and reached Surgut in 1975, has special significance. First, the tenth five-year-plan (1976–80) calls for the extension of this line as far again northwards to the Urengoy gasfield. This has great potentialities, in that it invites further extension to the lower Yenisey (Igarka, Dudinka), thus reaching the important mining area of Noril'sk, which is connected to Dudinka by a line built in 1937, converted to standard gauge in 1965 and electrified in 1975. Second, the Tyumen'-Surgut link is planned to be the beginning of a 'northern trans-Siberian', running parallel to the main line and up to 500 km north of it. East of Surgut it will follow the valley of the Ob' to Kolpashevo, thence to Belyy Yar, Lesosibirsk, Boguchany, Ust'-Ilimsk (all already linked, or soon to be linked, to the south), and, finally, following the route of the long-considered Baykul-Amur line (BAM), north of Baykal to Komsomol'sk-na-Amure. The next stretch east of Surgut as far as Nizhnevartovsk (216 km) was opened in 1976. Completion of the whole line will undoubtedly take many years, and approval has not yet been given to all stages of the project. But if completed, it will open up to further development a broad strip of territory immediately to the north of the present developed area.

Construction work started on the BAM in 1974, and may be largely completed by 1982. This section is 3,150 km long and runs from Ust'-Kut on the upper Lena to Komsomol'sk on the lower Amur. The original BAM included further sections at each end; from Tayshet on the Trans-Siberian to Ust'-Kut, and from Komsomol'sk to Sovetskaya Gavan' on the Pacific, thus forming a link between the Trans-Siberian in central Siberia and the sea. These two sections were completed in 1945 and 1950 respectively. From Tynda, near the midway point, a link with the Trans-Siberian at a station called Bam was completed in 1975. The terrain to be crossed by this long line includes some very difficult stretches – mountains, swamps, permafrost, areas subject to seismic disturbance and avalanche danger – and the project is reckoned to be the biggest and most complex now under way in the USSR. The line will be single-track at first, but with provision from the start for conversion to double-track. It will be electrified initially from its western end up to the Severo-Muyskiy tunnel northeast of Baykal, but ultimately the whole line will be electrified. At two points in particular there are major mineral deposits close to the proposed route: copper at Chara, in the Udokan region in the north of Chitinskaya Oblast', and both iron ore and high-grade coking coal near Chul'man in southern Yakutskaya ASSR. The Chul'man deposits have often been referred to as a possible basis for major industry in the Soviet Far East. A spur from Tynda on the BAM to Berkakit, near Chul'man, was due for completion in 1977 and will allow exploitation of the coal (an agreement was signed in 1974 with a Japanese company which will provide credits in exchange for some of the coal). The iron, which is farther north, will be reached by a further extension, approved in 1976. Thus this spur might then become a first stage in bringing the railway to Yakutsk.

Another line under consideration would run from Salekhard at the mouth of the Ob' to Igarka on the Yenisey, about 1,250 km. Construction started on this line in Stalin's lifetime (when it would have had little justification), but it was abandoned at his death. Oil and gas discoveries in the area bring the idea to life again, especially as the route passes through Nadym, which is due to become the transmission terminal for vast gas production. It was reported that a 150-km section of this line, from Nadym to the gasfield at Medvezh'ye, was put into use in 1974. The terrain is very difficult and wholly within the permafrost zone, but some 600 km of bed were prepared, and many bridges built. If more of this line were completed, it is likely that a bridge would be built over the Ob' at its mouth to make connection to the national network at Labytnangi. Thus either this line, or the proposed Surgut-Urengoy line could become the means of access to the remote Taymyr region, with its rich and increasingly profitable mining developments, and so of great importance for north central Siberia.

All the plans and projects just mentioned have either been completed, or are in construction, or are at a detailed planning stage. Other projects have sometimes been aired in the press. One is a third possible link with Noril'sk – a north–south line running parallel to the Yenisey river and meeting a spur of the Trans-Siberian at Lesosibirsk (Maklakovo). This was mentioned in 1974 as a general possibility, and a rather more specific proposal was advanced in 1975 for the construction of the most northerly section (about a quarter of the whole), from Noril'sk to Karasino, a point some 80 km upstream from Igarka. A second, more spectacular, is a western equivalent to BAM, running from the Yenisey at Lesosibirsk northwestwards to Surgut (this stretch would be a section of the 'northern Trans-Siberian' mentioned above), across the Urals, and on to the Barents Sea at Indigskaya Guba (not, however, using either of the existing spurs of the Vorkuta line). A third, most spectacular of all, would be a line to Bering Strait (which would at least begin to make sense of the proposals for bridging or damming that strait). But ideas of this sort must be sharply distinguished from projects on the planning of which much money is being spent.

The most severe environmental effect on road and railway construction is that produced by the presence of permafrost. Although track-beds and bridges are not in themselves generators of heat, which would then be conducted into the frozen soil, any disturbance of the surface is likely to upset the permafrost table and so lead to effects felt on the surface. A railway is more vulnerable than a road, because small movements in the track-bed are likely to have more serious effects on traffic movement. It is known that the northern section of the Vorkuta line faced serious problems in the early stages of its operation. It is highly likely that the Dudinka-Noril'sk line gave similar trouble. The course to be followed by BAM through the mountains northeast of Baykal lies in continuous permafrost, and will be a problem. Other lines so far constructed in the north extend only into areas of sporadic permafrost, so the problems are not likely to have been so severe. But much difficulty may be expected in construction of the Surgut-Urengoy line, and any extension northeastwards, and this may indeed be one of the factors delaying a definite decision on whether to complete the earlier work done on the Salekhard-Igarka line.

National significance

The Soviet government has always had the strongest motivation to develop its northern territory. From its earliest years, that government faced an international situation which dictated that the Soviet state could become strong and independent only if it could exploit, and where necessary find, the natural resources of its own territory. This was the stimulus behind the early emphasis on the northern sea route, the development of other transport media, and the intensive exploration for minerals. That exploration yielded some very important finds: new gold-bearing regions in the northeast, nickel in Kola and Noril'sk, and, most recently, diamonds in the Lena basin. All these, as already mentioned, occupy a significant place in the national economy. The proportion of northern output to national output is estimated to have been of the following order in 1970: gold, 50 per cent; nickel, 65 per cent; tin, 50 per cent; apatite, 100 per cent; diamonds, 100 per cent; iron, 4 per cent. Further, Soviet gold and diamonds have a significant place in world output, and Soviet production of both comes onto the world market. The same is becoming true of the oil and gas from the immense deposits of northwest Siberia. The oil and the gas are expected to provide, respectively, 50 and 35 per cent of Soviet output by 1980. The whole of the increase in oil production called for by the tenth five-year-plan (1976–80), and almost all the increase in gas production, is due to come from western Siberia. Gas is already being exported to European countries; it may go to the USA; and both gas and oil will probably go to Japan.

Mineral production as significant as this can clearly justify a very large investment. There are two other factors, however, which are likely to have been taken into account as well. One is strategy. The word may be taken to comprehend strictly military planning, such as the construction and maintenance of military airfields, rocket-launching sites, radar stations, and naval bases, all of which are likely to exist in the north. It includes the provision of raw materials of strategic importance, and the north's nickel, tin, gold, and diamonds are obviously in this category. It could also be held to cover the exercise of sovereignty over remote regions susceptible to claim by others. At this point it becomes hard to distinguish strategic motivation from the motivation mentioned at the beginning of this section, the desire to be strong and self-sufficient. The two merge, and it is one of the strengths of the Soviet system of government that this situation causes no administrative difficulty. It is, of course, impossible to identify what proportion of the Soviet northern budget might in a western context be the responsibility of departments of defence. If the Soviet defence interest in the north is at all comparable to that of the USA (and it is hard to believe it can be otherwise), then one can only suppose that that proportion is a significant one.

The other factor which may, at a rather lower level of probability, have played a part in justifying northern development in the Soviet mind is the idea of regional self-sufficiency. At various times during the Soviet period emphasis has been placed on the desirability of major regions of the country aiming at self-sufficiency. This has never applied to the north as a single region, but an argument in favour of increasing the self-sufficiency of, say, eastern Siberia, could have large implications on resource development in the northern part of that region. This motivation is essentially a compound of economic and strategic factors.

For all these reasons, therefore, it is clear that the north is very important to the USSR, economically and strategically, and that the investment in it is such that its importance is unlikely to diminish. Indeed, expansion is foreseen at a greatly accelerated rate. Population of 'the northern regions', not exactly defined, is expected to grow by a factor of four in the twenty years 1972–92, according to a statement from the Siberian division of the Academy of Sciences on 26 February 1973. If this is to become official policy, it is very remarkable, and calls for further examination. The population of the major administrative areas in the north of the USSR (listed in Appendix 1), probably a rather smaller area than the undefined 'northern regions', increased in population by a factor of two over thirty-three years (1939–72). The increase in the sixteen years 1959–75 was by under 40 per cent. The 'far north' of Soviet statistical handbooks, an area which probably approximates rather closely to 'the northern regions', increased by 45 per cent in thirteen years (1961–74) – a somewhat faster rate. But this is still very much slower than what will be required for the suggested increase over the next twenty years. For that, the rate achieved in the last decade would have to be six times as great – an increase so vast, and with such huge implications for housing, transport, and other social facilities, as to appear almost absurd. In addition, Siberia as a whole has been experiencing a net out-migration in recent years. Of course, the target figure may be lowered as time passes, but even if the reality were to be half what is now in mind, the increase would be very large. The significant thing is that serious thought could, in 1973, apparently be given to an increase of these dimensions.

A considerable part of the increase will no doubt result from the oil and gas development in northwest Siberia. Here there is a resource of very large dimensions, and the population of the oil- and gasfields may well quadruple in the next twenty years. Production at Noril'sk is expected to triple by 1980, exploiting the new deposit at Talnakh. But it is quite clear that not only these areas were in mind. The whole of the north is given great importance by planners. The chairman of Gosplan's Council for the Study of Productive Forces of the USSR, Academician N. N. Nekrasov, wrote in 1971: 'it is now quite clear that the future of our productive forces as a whole depends largely on the organization and pace of development of our northern territory.'

CHAPTER THREE

Canada: the slow retreat of 'the north'

Geographical background

In the years following Confederation, Canada expanded west across the continent and north to the Arctic Ocean, acquiring Rupert's Land and the North-western Territory from the Hudson's Bay Company and arranging with the United Kingdom for the transfer of the northern islands to the new Dominion. Virtually all the Canadian northland formed part of this enormous area and was included within the North-West Territories, but the government did little to administer it. Some of it remained uninhabited and little explored, and the rest was left in the hands of the fur trade and the missions. The subsequent political development of Canada, which entailed the establishment of new provinces, the northward expansion of provincial boundaries, and the parthenogenesis of a new territory, were at the expense of the North-West Territories, out of which great slices were cut. In 1912 the present political boundaries were established, apart from that between Quebec and the then Dominion of Newfoundland which was decided by the Imperial Privy Council in 1927, and the inclusion of Newfoundland as a Province of Canada in 1949.

The political divisions took little account of geographical considerations, and the Canadian northland falls under several different jurisdictions. In the east, Newfoundland and Quebec extend north up to the shores of Hudson Strait and Ontario to Hudson Bay. In the west, Manitoba, Saskatchewan, Alberta and British Columbia, reach the 60th parallel. Seven provinces, therefore, include parts of the northland; farther north

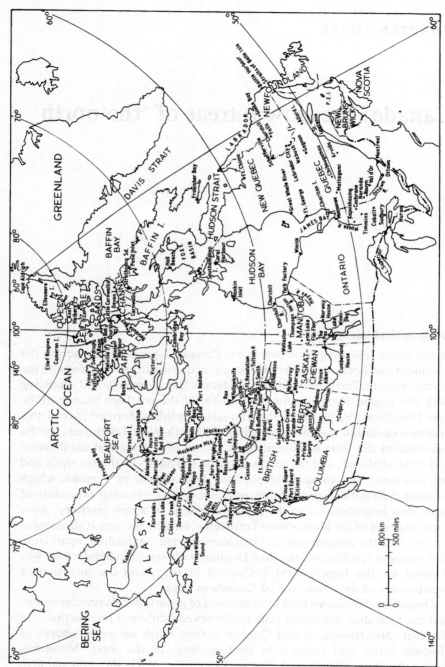

Map 10. Canada

and wholly within it lie the two territories, the Yukon Territory and the residual part of the Northwest Territories which has lost its hyphen as well as much of its land. The Northwest Territories has three districts – Mackenzie, Keewatin, and Franklin, but they are rarely used as political divisions and their boundaries have little more significance than lines on a map. The northland is being continuously eroded as development in the provinces moves north, integrating the economy with that of southern Canada.

Canada has more northern territory than any other country except the USSR. If just the area north of the treeline is considered – the area that most closely meets the popular concept of 'Arctic' – Canada has the largest area, greater even than that of the USSR, where the forests extend much farther north. Geographically, Canada is certainly a northern country. Economically, culturally, and politically, however, Canada reflects the attitudes of her population, the great majority of which lives within 200 miles (320 km) of the east–west border with the United States, and whose way of life, like that of the United States to the south, has its roots in Europe.

The northland of Canada includes the Yukon and Northwest Territories and the northern parts of all the provinces except Nova Scotia, New Brunswick, and Prince Edward Island. It extends over approximately 2,750,000 sq miles (7 million sq km) of Canada, or nearly three-quarters of the whole country, and of this about one million sq miles (2.6 million sq km) lie north of the treeline. The southern limit is in Quebec where it reaches almost to latitude 48° N, less than 200 miles (320 km) from Montreal. From there to Cape Aldrich, the most northern point in Canada, is nearly 2,500 miles (4,000 km). From the Alaska–Yukon border in the west to Battle Harbour on the Labrador coast in the east is some 3,000 miles (4,800 km). The Canadian mainland extends just beyond the Arctic Circle in the west, but Hudson Bay penetrates deeply into the continent in the northeast. The Canadian Arctic Archipelago, north of its mainland, includes six of the thirty largest islands in the world.

The land

The dominant geological feature is the Canadian shield which lies exposed in a great incomplete ring of granites and gneisses around Hudson Bay and forms about two-thirds of the Canadian northland. The rocks of the shield were subjected to several periods of folding during Precambrian times, but the mountains then formed have long been worn away to a peneplain. It is rough country, with innumerable lakes, but rarely more than 2,000 ft (600 m) in altitude, except in the east. Here the shield has been tilted to form a mountainous fringe along the coasts of the North Atlantic, Davis Strait, and Baffin Bay. In some areas, in particular the

Hudson Bay lowlands along the southwest coast of Hudson Bay and on the more southern islands of the Archipelago, the shield is overlain by flat-lying Paleozoic limestones and sandstones which form lowlands and plateaus. Elsewhere these formations have been eroded away.

West of the shield lie the northern interior plains. Here the shield dips at a rate of 15 feet a mile (2.5 m per km) under horizontal Paleozoic strata which increase in thickness to the west to reach as much as 10,000 ft (3,000 m). This area forms most of the drainage basin of the Mackenzie River which runs northwest to empty through the many meandering channels of the Mackenzie Delta into the Arctic Ocean, and is the only navigable river in North America that flows into the Arctic Ocean.

Between the interior plains and the Pacific Ocean the Paleozoic strata have been uplifted and folded to form the mountain ranges and plateaus of the Cordillera, through which the Yukon River and its tributaries have cut deep valleys. In southwest Yukon the St Elias Mountains rise to over 19,000 ft (5,800 m) and include Mount Logan (19,850 ft), the highest mountain in Canada.

North of the Canadian shield the Innuitian region covers most of the Queen Elizabeth Islands, the islands north of the Parry Channel which runs east–west across the Arctic Archipelago. Paleozoic, Mesozoic, and Tertiary formations have been laid down and folded repeatedly, resulting in sedimentary strata up to 60,000 ft (18,000 m) thick. Ellesmere, Axel Heiberg, and Devon Islands are extensively ice covered. The greatest heights are in northern Ellesmere, where the highest mountain is about 8,500 ft (2,600 m), rather more than any in Baffin Island, and is in fact the highest in North America east of the Rockies.

The most recent formations lie along the Arctic shores of Banks and Prince Patrick Islands. Here beds of Tertiary and Pleistocene sand and gravel form a low-lying coast and probably extend far out on the Arctic continental shelf.

In comparatively recent times the northern landscapes have been re-fashioned as a result of the Pleistocene glaciations. Glaciers cut great U-shaped valleys through the mountains, and the advancing ice sheets scoured and scarred the land. As the ice melted, masses of glacial till were left as erratic boulders, moraines, drumlins, and eskers. This has given a 'grain' to the country, leading from the old centres of glaciation. Travelling across the grain is much more difficult than travelling with it.

The ice also changed the sea level. As water was withdrawn from the oceans to form the inland ice, the level of the sea sank. However, the weight of this great mass of ice depressed the land and it too sank, more slowly but farther. When the ice melted these processes were reversed. This has left its mark on the land, where old shore lines and raised beaches can often be found many miles inland and at heights up to 750 ft

(225 m). The drainage pattern was also affected, with the rivers and streams rejuvenated by the uplift. In some parts of the north the land is still rising at rates that may be up to a metre in a century.

In most of the northland the ground remains frozen below the surface throughout the summer. This hinders drainage and much of the melted snow remains in shallow lakes and swampy ground. The effects of permafrost can also be seen where cracks in the surface of the ground have been deepened by repeated freezing and thawing to form distinctive polygons or circles from a few centimetres across to several metres. On slopes the thawed surface layer may slip downhill to produce long stripes. The most striking permafrost features are the pingos near the mouth of the Mackenzie. These conical hills, up to 50 m in height, and with cores of solid ice, are formed when water is extruded from the sites of old lakes as the permafrost reforms.

Climate

Most of the Canadian northland forms part of the mainland of the North American continent. Cut off from the modifying effect of the Pacific by the mountain ranges of the Cordillera, it experiences a northern continental climate with long winters and short, at times warm, summers. At Snag in the Yukon Territory, the temperature has fallen to −62 °C. This was exceptional, but −50 °C is not unusual. At the other extreme, 39 °C has been recorded at Fort Smith in the Northwest Territories.

Along the Arctic coast and in the islands to the north and east, temperatures are less extreme owing to the effect of the sea, but the mean annual temperature is lower. The winters are long, dark, and cold, with temperatures often falling below −45 °C for days on end. The sea ice melts in most areas during the long days between the spring and fall equinoxes, but the temperature rarely rises above 15 °C. There are frequent fogs, and summers as they are known in temperate regions do not exist. Frost and snow can occur any month of the year. The Western Canadian Arctic lies in an area of higher pressure and has calmer weather than the Eastern Arctic where the Davis Strait cyclonic zone brings frequent storms, especially in the fall. Precipitation is low, particularly north of the Arctic Circle where annual precipitation averages less than 10 inches (25 cm).

Vegetation

The treeline in Canada runs approximately southeast from the Mackenzie Delta in the west to Churchill, skirts around James Bay, and then crosses northern Quebec to the Atlantic coast. It divides the Arctic from the sub-Arctic, the barrens from the bush, and it is a very real boundary. To the south lie the boreal forests where the snow lies soft and deep in the winter. The trees are mainly conifers, with white spruce characteristic. Tamarack and black spruce are common in the north and east, jackpine in the west, and lodgepole pine and alpine fir in the northwest. Deciduous trees, particularly white birch and poplars, become more frequent towards the south. This is the country of the Indians, of snowshoes, canoes, and toboggans. There is wood for fuel and for building cabins. To the north stretches the tundra. Here there is no shelter. When the wind rises the snow is driven across the land in blizzards in which a man cannot see more than a few feet and the snow is packed into hard drifts. It is the country of the Eskimos, of sledges and snow houses. Arctic willow, dwarf birch, and other low-lying shrubs, and blubber from sea mammals are the only sources of fuel. Though the tundra appears white and barren in the winter, it is transformed when the snow melts in the long days of May and June and flowering shrubs and plants, grass, mosses, and lichens make it a land of variety and contrast.

Near the treeline, but rather to the south of it west of Hudson Bay and to the north of it east of Hudson Bay, is the southern limit of continuous permafrost, the area where the temperature of the ground beneath the surface remains below 0 °C throughout the year. South of this line there is a band where patches of permafrost may occur. The presence of permafrost causes a number of problems which increase the costs of operating in the north.

The people

The aboriginal population of the northland was Indian and Eskimo. The northern Indians were essentially an inland people whose life was adapted to the forests. They belonged to one of two linguistic groups, the Athapaskan to the west and north, and the Algonkian to the east and south. In contrast the Eskimo, or Inuit as they now frequently prefer to be called, were a maritime people living on the Arctic shores of the tundra, and sharing with the Greenlanders and the North Alaskan Eskimos the Inupik dialect of the Eskimo language. Both races were hunters, living in small scattered groups and, despite the immense area they covered, their population in the Canadian northland was probably never more than around 40,000, more or less evenly divided between Eskimo and Indian. The resources of the northland and the hunting techniques that had evolved could not support a larger population. Starvation was the control that limited their numbers. The treeline formed the boundary between the two races and here they met from time to time. Each distrusted what it did not understand and there is a history of fear and hostility between them.

The animals

There are not many species of land mammals north of the treeline. The most important economically are the barren ground caribou, which is hunted by the Eskimos for meat and skins, and the Arctic fox, the only significant fur-bearer. The Arctic fox preys on the lemmings and both follow a cycle usually lasting four years during which the population builds up and then collapses. Polar bear, muskox, and Arctic hare are also of some value either as skins or food. The few other species, such as wolf, wolverine, and ermine, have probably never contributed significantly to the economy.

Ringed and bearded seals are found along all the northern coasts and walrus occur throughout the Eastern Arctic; they are the most important sea mammals, supplying the Eskimos with food, skins, and oil. Harp and hooded seals are migratory, with a range extending along the coast of Davis Strait. Harbour seal occur in the Mackenzie Bay as well as the Eastern Arctic. The larger whales are now rare but white whale are hunted throughout the north and narwhal in northern Baffin Island and Foxe Basin. Arctic char are much the most important fish north of the treeline and large numbers can be caught when they run up the rivers in the fall. In the spring flocks of birds migrate to breed in the north and many waterfowl are killed by the Eskimos. Ptarmigan are resident and are the only bird of any use as food in the winter.

The fauna is much more varied in the boreal forests than in the barrens to the north. Moose and woodland caribou are the most important animals for food, and beaver, muskrat, and mink for fur. Dall's sheep and mountain goat are hunted in the Yukon. Wood Buffalo National Park holds the largest existing herd of American bison, which numbers nearly 10,000. In the lakes and rivers white fish, trout, inconnu, herring, grayling, and pike are most frequently caught and, in the Yukon and the rivers of northern British Columbia, Pacific salmon. In some areas, especially James Bay, the Indians kill waterfowl on their migration routes. In Labrador Atlantic salmon are found in the rivers, and fishing for cod has long been an important occupation along the coast.

History

The geographical dichotomy in the north between the bush and the barrens is paralleled in the early history of the north. South of the treeline the fur trade provides the main historical theme; north of the treeline it is the search for the northwest passage.

The animals sought by the fur trade, the marten, mink, fisher, otter, fox, ermine, lynx, and, by far the most important, the beaver, lived in the boreal forests, through which the great fur-trading companies penetrated to the west and north. The Hudson's Bay Company developed their main supply route by sea through Hudson Strait to their posts in Hudson Bay, especially York Factory. From York Factory freight was sent, first by canoe but later by York boat, up the Hayes River to Norway House at the north end of Lake Winnipeg, and from there west to Cumberland House. Their rivals, the Montreal fur traders, followed a more difficult route. From Montreal they went by canoe up the Ottawa River and down the French River, across Lake Huron and Lake Superior, to Grand Portage and then northwest to Cumberland House. By the time the Hudson's Bay Company and the North West Company amalgamated in 1821, the fur traders had made their way along the rivers and lakes through the forests as far as the mouth of the Mackenzie. Their posts dotted the main river transport routes of the sub-Arctic, and in the following years they spread into the Yukon. These posts became the local centres to which the Indians brought their fur to trade and where the Roman Catholic and Protestant churches established their missions.

The Hudson's Bay Company, its economic pre-eminence strengthened by its charter, was the power in the land. Even after 1869, when the Hudson's Bay Company surrendered its charter to Rupert's Land, the situation in the north changed little. The future of Canada was seen as lying in the west, and the government concentrated its efforts on the development of the prairies and building the east to west communications across the country. The Hudson's Bay Company continued to dominate the northern fur trade.

It was in the west that the pre-eminence of the fur trade in the economy of the sub-Arctic was first challenged. In the second half of the nineteenth century placer-gold discoveries had drawn prospectors to northern British Columbia and from there into what later became the Yukon Territory. In 1898 the discovery in Bonanza Creek of exceptionally rich placer gold led to the Klondike gold rush, bringing thousands of men from southern Canada, the United States, and elsewhere to the north. By the end of the century, Dawson City had become the largest city in Canada west of Winnipeg, with a population of over 25,000, considerably greater than that of the whole Yukon Territory today.

Elsewhere the encroachment on the dominance of the fur trade was much more gradual. The extension of rail across the continent provided a more economical way than water transportation of sending supplies to the west, and the Hudson's Bay Company developed routes leading north from the railways, using stern-wheeled paddle steamers where the rivers and volume of freight allowed. The most important of these arteries led north from Edmonton by an overland trail for 100 miles (160 km) to Athabasca Landing and from there down the Athabasca, Slave, and Mackenzie Rivers to the Arctic Ocean. The improved transportation brought white trappers and independent traders into the country and commercial fishing began in many of the lakes, spreading from Lake Winnipeg in the 1870s to reach northern Saskatchewan by the 1890s. The religious missions, fiercely competitive with one another, proselytized the Indians and operated schools and hospitals with the aid of limited government subsidies. The government restricted its own activities mainly to exploratory surveys carried out through the Geological Survey of Canada.

In 1888 a Senate Committee on the Resources of the Great Mackenzie Basin was appointed and it published a highly enthusiastic report on the potential for agriculture, fisheries, forestry, mining, and petroleum, setting the precedent for the optimistic and promotional tone that has continued to this day to pervade government pronouncements on northern resources. The Senate Committee report had little success in speeding development, but it was followed by an enlarged program of government surveys.

In the eastern Canadian sub-Arctic the agents of change were the mining and forestry industries. Nickel and copper ore had been discovered at Sudbury in 1883 during the construction of the Canadian Pacific Railway, but it was another chance discovery as a result of railway construction, that of silver in 1903 at Cobalt when the Northern Ontario Railway was being built, that led to the recognition of the mineral potential of the Precambrian shield and the development of gold-mining in northeastern Ontario.

The lumber industry was centred on pine and scarcely touched the northland but the last years of the nineteenth century saw the beginning of the pulp and paper industry which depended on spruce. The industry spread rapidly into the forests of the north where many of the rivers could be developed to provide hydroelectric power. This penetration into the north has continued to the present day, the pulp industry moving farther north and west in its search for spruce, and mines being developed as the price of metals rises enough to cover increasing operating and transportation costs of mining in more distant areas and sometimes more difficult terrain.

The fur trade had little interest in the land north of the treeline. Here it was the search for the northwest passage to the Orient that led to the exploration of the Arctic coast of the continent and the islands to the north. Frobisher led the way in 1576 when he reached Baffin Island, and he was followed during the seventeenth and eighteenth centuries by other adventurers seeking a passage to the west but finding their way blocked by land or impenetrable ice. After the conclusion of the Napoleonic Wars the quest was taken up by the Royal Navy and reached its climax in the middle of the nineteenth century in the many searches for the lost Franklin expedition, during which the north coast of the continent and most of the islands of the Arctic Archipelago were mapped. Sverdrup and Stefansson completed the major discoveries but some areas remained unexplored for many years. The last discoveries were not made until after the Second World War when air photographic missions revealed hitherto unknown islands in Foxe Basin.

The ships of the Hudson's Bay Company en route to York Factory traded sporadically with the Eskimos in Hudson Strait during the eighteenth and nineteenth centuries, but the Company did not reach out into Eskimo country and was in fact criticized for this lack of enterprise. Sometimes the Company sent a ship north along the west coast of Hudson Bay to trade but the Eskimos often had to make long journeys to the Company's posts. Except on the east coast of Hudson Bay it was not until well into the twentieth century that the Hudson's Bay Company established its first trading post north of the treeline. It was left to the whalers to follow in the tracks of the explorers and sometimes to be ahead of them. By the seventeenth century whaling had reached Davis Strait, but it was restricted to the Greenland coast until 1818 when whalers found their way through the North Water in Baffin Bay to the rich whaling grounds off the north and east coasts of Baffin Island.

Apart from one or two unsuccessful attempts by the Hudson's Bay Company in the eighteenth century, commercial whaling did not begin in Hudson Bay until 1860 when the first two whaling ships passed through Hudson Strait. Their success brought more whalers to the northwestern part of Hudson Bay, and the next half-century saw the whale stocks depleted. At first the whalers were predominantly American, but British whalers played a more important part during the years when the industry was declining. As the whaling ships usually spent two seasons in Hudson Bay and wintered there, they had extensive contact with the Eskimos who hunted for fresh meat for the crews and assisted in the whaling. In this way whale-boats, guns, and other goods were introduced. When whales became scarce towards the end of the century, the whalers turned increasingly to walrus, muskoxen, foxes, polar bear, and seals. In 1902 a whaling company opened a small mica mine at Lake Harbour which produced between 10 and 20 tons of mica annually for several years. However, attempts to diversify could not compensate for the decrease in the number of whales and the introduction of cheaper substitutes for whale oil and baleen, and the last whaling ship left Hudson Bay in 1915. The same factors led to the decline of the industry along the east coast of Baffin Island. Whaling had an even shorter history in the Western Canadian Arctic. The first whalers reached Herschel Island in 1888 and within twenty-five years the industry had run its course.

In both the Eastern and Western Canadian Arctic the Eskimos had come to depend on trade goods, and the withdrawal of the whalers would have left them with no source of supply if the fur trade had not begun at this time to expand into the north, filling the vacuum left by the whalers. In the succeeding years the many smaller trading companies were gradually taken over by the Hudson's Bay Company, which by the Second World War had extended its network of posts to cover all the territory occupied by the Eskimos.

In Labrador a flourishing Basque whale fishery based on the deep-water harbours on the north side of the Straits of Belle Isle began as early as the first half of the sixteenth century and was producing about 5,000 tons of oil in at least some years. It declined rapidly towards the end of the century probably owing to the demands for ships and men for the Spanish Armada and the losses incurred in its defeat. On at least two occasions the Basques wintered on the coast and were certainly in contact with the native people but do not appear to have traded extensively with them. The Basque whalers withdrew from Labrador probably in the first years of the seventeenth century. For the next two hundred years a few fishermen and fur traders formed the permanent European population.

In northern Labrador the most important influence was the evangelical Moravian Church, which established a number of missions towards the end of the eighteenth century and assumed control of the economic as well as the spiritual life of the Eskimos. During the nineteenth century the white population increased in southern Labrador, taking up trapping and hunting as well as fishing, and the Eskimos declined in number or withdrew to the Moravian settlements to the north. In the interior of Labrador and New Quebec the Montagnais and Naskapi Indians became trappers, taking their fur to trading posts on the north shore of the St Lawrence and the coast of Labrador. This pattern persisted with little change until the outbreak of the Second World War.

The situation at the outbreak of the Second World War

Throughout the north the Second World War was a turning point. Up to the outbreak of the war, development in the northern territories had proceeded gradually and in some areas scarcely at all. With the exception of the Klondike gold rush, the advent of the whalers, and some intensive mineral prospecting in the Great Bear Lake and Great Slave Lake areas, there had been no sudden major bursts of activity and an Indian, an Eskimo, or a fur trader from the early nineteenth century would not have felt out of place a hundred years later. Even this slow pace of development was halted during the first two years of the war as men and materials in Canada were diverted to supporting the war effort, and the war was far away. A dramatic change followed after the United States entered the war. It is convenient, therefore, to take stock of the situation as it existed at the beginning of the war before looking at the changes that have occurred since.

The Yukon Territory

In the Yukon Territory the Klondike gold rush had quickly faded, but it left behind two important legacies. One was the White Pass and Yukon Railway from tidewater at Skagway to Whitehorse, which was begun in 1898 and completed two years later. It immediately became the main transport route to the Yukon, supplies being sent to Whitehorse and then along the Yukon River. No longer was it necessary to follow the very much longer route to the Bering Sea and up the Yukon. The other legacy was the Yukon Territorial government which consisted of a fully elected council responsible for legislation and a commissioner, appointed by the federal government, with executive powers and directing a territorial public service. It was a form of government designed to respond to the needs of a rapidly developing area, and was left like a fish on the beach as the tide of development ebbed.

The rich gold placers that had led to the gold rush were worked out within a few years. Gold-mining survived, largely because of rationalization which allowed the use of hydraulic plants and dredges, but at a very much lower level. The economic collapse of 1929 helped gold-mining and the increase in the price of gold, which reached $35 an ounce in 1935, was another stimulant, but by 1939 annual production had risen to only just over $3 million, almost all by a single company, Yukon Consolidated Gold Corporation.

The gold rush had led to interest in other minerals in the Yukon, but mining was faced with very high transport costs. Copper was produced at a number of small mines in the Whitehorse area for some years and during the First World War annual production rose to over 1,000 short tons (900 tonnes), but the price of the metal fell after the war and the mines had to close. The other metals mined were silver and lead. Placer-gold mining had produced some silver, and early in the twentieth century a number of pockets of rich silver ore were exploited in the southern Yukon. In 1906 silver-lead ore was discovered in the Mayo area and production began in 1914. Additional ore was discovered after the war and a concentrating mill, built in 1925, operated until 1932 when the ore was depleted. However, new ore was again found, and by 1939 the value of production of silver and lead was nearly $2 million and would have been higher but for shortage of transport on the Yukon River. A small amount of coal was mined in the territory for local use.

Apart from trapping for fur, of which the annual value in the years preceding the Second World War was about $300,000, renewable resources added little to the economy. Attempts to introduce fur farming had not proved successful and only a handful of fur farms remained. The Territory held no prospect for significant commercial fishing. The gold rush had encouraged farming, and a number of cattle were raised and vegetables grown to meet the local market, but production and interest declined with the population.

By the beginning of the Second World War the total population of the Yukon Territory had fallen below 5,000, of which Indians formed about 30 per cent. For twenty years Dawson had been a city of only 1,000. In the Yukon, the past dominated the future.

The Mackenzie Basin

The routes that had supported the Klondike gold rush and the subsequent development of the Yukon led mainly from the Pacific. Elsewhere in the western northland, development spread out from Edmonton. The Grand Trunk Pacific Railway to Prince Rupert opened up much of northern British Columbia to forestry and prospecting. The railways to Peace River and Dawson Creek served the rich agricultural country of the Peace River where settlement had begun around the turn of the century and expanded rapidly in the late 1920s, and the line to Waterways gave access to the Athabasca oil sands and greatly improved the Mackenzie River transport route. It was along this last route that economic development penetrated deeply into the north. Continuing attempts to extract oil economically from the Athabasca oil sands proved unsuccessful, but in 1920 the Imperial Oil Company struck oil at Norman Wells less than a hundred miles (160 km) south of the Arctic Circle, and this initiated a short-lived surge of activity in petroleum exploration along the valley. Further exploration at Norman Wells failed to live up to the initial high hopes for a major oilfield. A small still was operated for a year to provide gasoline and kerosene for local needs, but the wells were capped in 1925. They were reopened in 1932 to provide for the mining activities on Great Bear and Great Slave Lakes and by 1939 annual production of petroleum had reached 20,000 barrels (2,750 tonnes).

Improved transportation, especially the use of float-equipped aircraft, encouraged prospecting in the Mackenzie District. The first mine to be developed was the Eldorado Mine at Port Radium on Great Bear Lake where silver-copper and silver-pitchblende ores had been discovered in 1928. A concentrator was built and production began in 1933, reaching by 1939 an annual value of nearly $2.5 million, much the greater part from the pitchblende. Meanwhile, in the Yellowknife Bay area of Great Slave Lake, where miners bound for the Klondike at the end of the nineteenth century had reported gold, renewed prospecting led to the development of a number of mines, and the construction of a hydroelectric plant. The first gold brick was poured in 1938 at the Consolidated Mining and Smelting Company's mine and production in the Yellowknife area was approaching $2 million in 1939.

A gold mine and hydroelectric power plant were developed at Goldfields in northern Saskatchewan on the north shore of Lake Athabasca. Production began in 1939 but the ore grade proved much lower than had been anticipated and the mine closed within three years. Lake Athabasca was also the scene of a commercial fishery which had then been operating for many years. Though the exploitation of the Athabasca oil sands remained for the future, a salt mine at Waterways in the same area was producing several thousand tons of salt a year by 1939.

Apart from farming in the Peace River and some limited forestry in the southern fringe of the northland, trapping and hunting were the only activities based on renewable resources that contributed significantly to the economy, providing a livelihood to the native population and to the white trappers who had moved into the north. Some vegetables were grown successfully for local use at posts along the river, and some timber was cut for the mines. An innovation was the introduction in 1933 of a reindeer herd from Alaska to the Mackenzie Delta area in an attempt to establish a new occupation for the Eskimos, as had been done in Alaska.

The volume of freight along the Mackenzie River was greatly increased by the mining developments in the Mackenzie Valley, and to this was added the supply of the Western Arctic. Freight for the Western Canadian Arctic had been sent by ship around Alaska to Herschel Island for distribution from there to the posts to the east, but ice conditions and lack of harbours along the north coast of Alaska made it a difficult and dangerous route. In both 1926 and 1931 the supply ship was lost and in 1933 it failed to reach Herschel Island. The Hudson's Bay Company therefore decided to send their supplies by barge down the Mackenzie River route and to use Tuktoyaktuk for trans-shipment to coastal shipping.

The increased freight along the river brought competition to the Mackenzie River Transportation Company, the transportation subsidiary of the Hudson's Bay Company which, like its parent, had enjoyed a virtual monopoly in the north. Its most serious competitor became the Northern Transportation Company, controlled by Eldorado Mines. By 1939 freight leaving Waterways had reached some 20,000 tons, of which the Northern Transportation Company was carrying about one-third, and it also handled nearly three-quarters of the much smaller southbound freight. Competition and increased tonnage greatly reduced rates. Freight from Waterways to Aklavik cost $12 per 100 lb (45 kg) in 1930 and to Great Bear Lake $16 in 1932; in 1938 the rates were down to $6.00 and $5.50 respectively. More efficient equipment was introduced, with diesel tugs and barges replacing the stern-wheeled paddle steamers and scows. Meanwhile overland trails were being developed north from the railways, and in 1939 the first tractor train reached Hay River from Grimshaw, and passed through to Yellowknife.

Even more dramatic was the growth of air transport. The Norman Wells oil discovery in 1920 led the following year to pioneer flights down the Mackenzie Valley. During the next few years, aircraft were used only occasionally but in 1927 the Royal Canadian Air Force began air photography for mapping, basing their aircraft at first at Fort Smith, and initiating a program that was continued and expanded up to the war. Within the next year or two aircraft added a new dimension to northern life, overcoming the isolation and remoteness of the sub-Arctic forests. Prospectors could reach areas that had been inaccessible to them, supplies could be sent in, and inspection trips that had taken months could be completed in days. Early in 1929 a passenger and airmail service was initiated to Fort Resolution and Fort Simpson, and it was extended all the way down the river to Aklavik by the end of the year. The depression slowed down this expansion, but in the mid-1930s it resumed. By the outbreak of the war, aircraft were being used by trappers to reach their trap-lines, for flying fish out of lakes in northern Alberta and Saskatchewan, for missionary and police work, and even for stealing fur from trading posts. Where there was enough traffic, and especially if an air mail contract could be secured, regular scheduled flights were introduced. In 1938 the three main companies in the Mackenzie Basin carried 1,300 tons of freight and 11,888 passengers, and the total air traffic was considerably greater.

Development required communications as well as transportation, and this need was met largely by the Royal Canadian Corps of Signals, which in 1923 had established radio stations at Dawson City and Mayo. A network of stations was set up in the 1920s, covering the Mackenzie District as well as the Yukon, handling meteorological reports and messages, and sometimes providing local broadcasts.

The first government representatives in the Northwest Territories were detachments of the North West Mounted Police established along the Mackenzie River early in the present century. At this time the comptroller of the NWMP was also the commissioner of the Northwest Territories, but the discovery of oil at Norman Wells in 1920 was followed by administrative changes. The deputy minister of the Department of the Interior became commissioner of the Northwest Territories, a council of four Ottawa civil servants was appointed the following year, and an administrative headquarters for the Mackenzie District was set up in Fort Smith. In the two other districts of the Northwest Territories administration remained in the hands of the Royal Canadian Mounted Police.

Northern Manitoba

In northern Manitoba the most important development was the construction of the Hudson Bay Railway, which had been begun in 1910 but was not completed to Churchill until 1929. Though its primary purpose was for shipping grain from the prairies to Europe, it proved a controlling factor in the development of mining in the region it passed through. In 1917 the Mandy Copper Mine at Schist Lake, 65 miles (115 km) north of The Pas, went into production but transport costs would allow only high-grade ore to be worked profitably, and it closed three years later. In the same general area a large copper-zinc ore body at Flin Flon on the Manitoba-Saskatchewan border was developed by the Hudson Bay Mining and Smelting Company, which installed a hydroelectric power generator at Island Falls on the Churchill River, and built concentrating, smelting, and electrolytic plants. This very large mine was served by a branch line from the Hudson Bay Railway and began production in 1930, in due course taking over the Mandy property. Another important copper mine, Sherritt Gordon, was developed at Sherridon, 45 miles (70 km) northeast of Flin Flon. It also was served by a branch line from the Hudson Bay Railway, used power from Island Falls, and had its ore processed by the Hudson Bay Mining and Smelting Company. Production began in 1931, was suspended the next year owing to low copper prices, but resumed in 1937.

Northern Ontario and Quebec

In northern Ontario and Quebec the mining and forestry industries had expanded greatly during the 1920s and were recovering from the setback of the depression, which had in fact stimulated gold-mining. Here the northland was retreating rapidly as new mining towns were established, as winter roads were followed by rail and all-weather roads, as hydroelectric power plants were built, and as timber limits moved north. Arvida, Noranda, Rouyn, Val d'Or, Cochrane, Kapuskasing and other new towns were integrating the area with the southern economy. In 1931 the Timiskaming and Northern Ontario Railway, now the Ontario Northland Railway, reached Moosonee, which became the distribution point for posts in the James Bay area and for Winisk and Great Whale River, replacing Charlton Island which had been supplied direct from the United Kingdom.

Hudson Bay and the Eastern Arctic

In the Keewatin and Franklin Districts of the Northwest Territories and the littoral of Hudson Bay under provincial jurisdictions, fur was the only product of significance that could be exchanged for imported goods. Survival depended on hunting; trapping provided guns, ammunition, and other hunting equipment that had become necessities and the few modest luxuries, particularly tea and tobacco, that could be afforded. Throughout the north the fur trade was recovering from the world depression. In the early 1930s the value of fur production in the two northern Territories, which had for some years been over $2 million annually, was halved. It rose somewhat throughout the decade but was only just over $1.5 million in 1939. Nowhere in Canada and few places in the world were more isolated than what remained of the old exclusive fur-trade empire. Supplies were brought in by ship, which called once during the summer. For the rest of the year the only contact was by radio, and most posts were without transmitters until two or three years before the Second World War. The few white people who went north, the fur traders, missionaries, and members of the Royal Canadian Mounted Police who represented the government, recognized that they had entered into a new society when they reached the north, an area which shared similar conditions, interests, and way of life very different from those elsewhere. The rest of the world was considered and always referred to as 'the outside'.

The Second World War

The effect of the Second World War was felt throughout the north and in many different ways. Fur prices rose and the value of fur production in the Territories began to exceed pre-depression levels. Gold-mining, however, declined owing to labour shortages, as gold had low priority after the United States entered the war, but production of base metals for military purposes was accelerated by mining the richest ore and neglecting development. The uranium ore on Great Bear Lake was a special case, acquiring a sudden secret importance. It was however the direct military activities that dominated and shaped the development of the north.

In the northwest the Japanese operations in the Aleutian Islands and the threat of further attacks, together with the need to deliver aircraft to the USSR, led to a number of major undertakings. The Northwest Staging Route was developed as a chain of airfields to Fairbanks, and the Alaska Highway was built to supply them and to provide an overland route to Alaska. The Caltel telegraph line erected along its route ensured reliable telecommunications with Alaska. The Mackenzie Valley became the main scene of the Canol Project to supply Alaska with oil from the oilfield at Norman Wells. This entailed additional drilling at Norman Wells and the construction of a pipeline from the oilfield across the Mackenzie Mountains to Whitehorse, subsidiary pipelines from Whitehorse to Skagway, Watson Lake, and Fairbanks, and a refinery at Whitehorse. To support the project it was necessary to build the Canol Road along the route of the pipeline, a network of winter roads and airfields throughout the Mackenzie Valley, and greatly improved water-transportation facilities along the Mackenzie River.

In the northeast the major project was the Crimson Staging Route for ferrying aircraft to Europe and for returning for treatment in North American hospitals some of the heavy casualties that were expected from the invasion of Europe. Airfields were built at The Pas, Churchill, Coral Harbour on Southampton Island, Frobisher Bay, Fort Chimo, and Goose Bay, as well as in Greenland, together with meteorological stations to support operations along the route. Modern technology arrived in one bound into an area where there was only a handful of white men and where the Eskimos were living as hunters and trappers with much of their pre-contact culture unchanged.

The immediate purpose of the wartime projects in the north ended when peace was restored, but they left a legacy in the form of an infrastructure that could be used for northern development. It was not perfect but it was, in many areas, better than literally nothing. The Alaska Highway, the telecommunications, and the airfields had an obvious continuing value. Though the main Canol Pipeline had no peacetime role, it had resulted in great improvements in transportation along the Mackenzie River. Possibly an even more important legacy was the recognition of the strategic significance of the north. The development of long-range aircraft meant that North America could be attacked from the north, a threat that nuclear weapons compounded. It became important to learn about the north, to provide a defence against this new threat, and to develop counter-threats. In the years immediately following the war a number of weather stations were established jointly by the United States and Canada in the Queen Elizabeth Islands to meet both military and civil needs, a chain of low-frequency Loran stations was built to assist navigation, an oil pipeline was constructed from Haines across northern British Columbia and the Yukon Territory to Fairbanks, and air photography and mapping were accelerated. These steps were followed by the Pinetree, Mid-Canada, and Distant Early Warning radar lines, Strategic Air Command air refuelling bases in Northern Alberta, Churchill, and Frobisher, and sophisticated communications required for the Ballistic Missile Early Warning System. The effect of these very major projects in providing employment for the native people is frequently exaggerated, but they added greatly to the infrastructure, which again was designed for the needs of defence rather than of northern development.

Resources and their development

Mining and hydroelectric development

The post-war years have seen a number of very large mines brought into production and hydroelectric plants installed in the northland, but they have been mainly in the northern parts of the provinces. The scale of these major developments has been much too great and the technology too advanced for the small and untrained northern population to participate significantly. Roads and railways from the south have penetrated the northland to these resources. Southern Canadians have moved in to exploit them and the northern economy has retreated farther. In the Territories themselves only the southern parts of the Yukon and the Mackenzie District have shared at all in this expansion. Elsewhere the north has remained largely a land of promise. Much has been written about its potential wealth and increasing efforts are being put into mineral exploration. Some discoveries have been made but they have not proved rich enough or large enough to be exploited economically at the present time.

The growth in the southern fringe of the northland has been very impressive. The harnessing of Churchill Falls to develop 5,225 MW in *Labrador*, and the Manicouagan River to develop 1,185 MW in *Quebec*, and the large-scale exploitation of iron ore on the Labrador-Quebec border near Schefferville, followed by other iron ore deposits to the south and west at Carrol Lake, Wabush, and Mount Reed, and titanium ore at Allard Lake, have transformed the north shore of the St Lawrence. Copper is being mined in the Chibougamau area, and copper, lead, and zinc in the Mattagami district. Farther north the giant James Bay hydroelectric development with a planned capacity of over 10,000 MW is now under way. The most northern development in Quebec is in northern Ungava, where an asbestos mine was brought into production in 1974 by the Asbestos Corporation, mining one million tons of ore a year for treatment in West Germany. In the same area the New Quebec Raglan Mine holds rich nickel-copper ore bodies containing over 16 million tons of ore grading 2.48 per cent nickel and 0.71 per cent copper. It is awaiting favourable conditions for development.

Several new ore bodies have been brought into production in northern *Ontario*, including the Texas Gulf Sulphur Company's very large silver-zinc deposit near Timmins. The major hydroelectric developments have been in the Moose River Basin.

In northern *Manitoba* the International Nickel Company has developed an enormous nickel mine near the Hudson Bay Railway and established the town of Thompson, which in a few years has grown to a population of over 17,000. Sherritt-Gordon closed down the Sherridon operation in

1950 but has developed other copper-zinc mines at Lynn Lake, Fox Lake, and Ruttan Lake in its place. More ore deposits have been brought into production near Flin Flon on the Saskatchewan border and copper concentrates are trucked there for smelting from as far as Walden Lake, a haul of 266 miles (426 km). Major hydroelectric plants have been installed on the Nelson River, and are being increased to generate over 3,000 MW.

Farther west in *Saskatchewan*, uranium-mining has become of great importance. Shortly after the war the development of the Ace uranium mine near Lake Athabasca by Eldorado Mining and Refining Company was followed by a number of other mines in the same area. Apart from the Ace mine, they closed down after a few years, but a mine at Rabbit Lake near Wollaston Lake has recently begun operations and other uranium ore bodies are being developed. A detailed examination of the hydroelectric potential of the Churchill River is being made.

In northern *Alberta* there is little metallic mineralization, but major hydroelectric generating plants are being considered for the Peace and Slave Rivers.

Northern *British Columbia* is rich in metals, and improved technology has allowed the development of several very large low-grade ore deposits, especially of molybdenum and copper. In this area too, the Aluminum Company of Canada has developed the Kitimat complex for the production of aluminium. The most northern major mine in British Columbia is the Cassiar asbestos mine. Fibre from this mine is trucked to the Alaska Highway and then north to Whitehorse for transport by rail to Skagway and on by sea to Vancouver. Although situated in northern British Columbia the Cassiar asbestos mine is more part of the economy of the Yukon Territory. Recent hydroelectric installations on the Peace River, developing over 2,500 MW, lowered water levels in the Mackenzie Basin in northern Alberta and the Northwest Territories, particularly while the reservoirs were filling, and damaged trapping in the Athabasca Delta in Lake Athabasca.

In the *Yukon* dredging for gold continued for many years but rising costs and the fixed price for gold inevitably brought it to an end, the last dredge closing down in 1966. A few small placer-gold properties are still worked, encouraged by the rise in the price of gold, and new rich placer-gold deposits have been reported near the Alaska border, south of Dawson.

United Keno Hill Mines has remained in operation, milling up to 400 short tons a day. The ore is rich in silver and about 3 million oz (93 tonnes) are recovered annually, together with some 7 million lb (3,000 tonnes) of lead, and roughly half that amount of zinc.

Prospecting, often by local individual prospectors, in the years following the Second World War brought a number of important mining possibilities to light. The Cassiar Asbestos Corporation found asbestos ore at Clinton Creek near Dawson, and developed a mine to produce some 100,000 short tons of asbestos fibre annually of grades complementary to those from its operation at Cassiar. Mining began in 1967, but the ore has proved to be less extensive than anticipated and the mine will probably close in 1978. The most important discovery so far has been a very large ore body of lead and zinc with some silver and gold situated 130 miles (200 km) northeast of Whitehorse, where the town of Faro has been built. It was developed by the Anvil Mining Corporation with substantial government assistance in the provision of infrastructure and supporting services. Production began in 1969 and 3 million short tons of ore a year are now milled to produce 370,000 short tons (340,000 tonnes) of concentrates for shipment to Japan, and 90,000 short tons (80,000 tonnes) for Germany. About 25,000 short tons of coal a year are mined and used to dry the concentrates. A third substantial operation in recent years has been the resumption of copper-mining in the Whitehorse area where Whitehorse Copper Mines produces 20 million pounds (9,000 tonnes) of copper annually.

Several other mines have operated for a few years in the Yukon Territory but have failed to survive. The total value of production has, however, increased steadily and in 1975 it exceeded that of the Northwest Territories (see Table 5) and reached 4 per cent of the Canadian total. A decrease in 1976, the first for many years, was the result of a prolonged strike at the Anvil Mine, much the largest producer. With mining for base metals now firmly established and with a number of promising prospects, the future of the industry in the Yukon seems bright. Its contribution to the national economy is limited by the fact that a large part of the output is exported in the form of concentrates. Employment in the producing mines totalled 1,335 in 1975.

Only a small part of the hydroelectric potential of the Yukon Territory has been developed. The Northern Canada Power Commission, which is responsible for providing and operating power plants in both the Yukon and the Northwest Territories, has developed 5 MW on the Mayo River, 20 MW, now being doubled, on the Yukon River at Whitehorse, and over 30 MW on the Aishihik River. Elsewhere in the Territory diesel engines are used to generate power. A very major development would be possible by diverting water from the Yukon River through the Coast Range to the Pacific. This was proposed early in the 1950s but has not been raised recently. Any diversion of the Yukon water would have international complications as the river flows through both Canada and Alaska.

In the Mackenzie District of the *Northwest Territories* the Eldorado Mine at Port Radium, which had been taken over by the Canadian government during the war, closed in 1960 when the ore was exhausted. The two mines in Yellowknife, Giant Yellowknife and Con-Rycon, have continued in operation, together producing about 200,000 oz (6 tonnes) of gold a year. Other mines in the area have closed down.

By far the most important development has been at Pine Point south of Great Slave Lake where lead-zinc ore had been reported by miners on their way to the Klondike at the end of the last century. The Consolidated Mining and Smelting Corporation decided to exploit this deposit, having first received an assurance that the government would finance most of the construction of a railway to the mine from the south. The railway was built by Canadian National Railways from Peace River and completed in 1964. Production at Pine Point has exceeded 200 million lb (90,000 tonnes) of lead annually and about double that amount of zinc. Over 60 per cent of the concentrates are shipped to the Cominco smelter at Trail in British Columbia, and the rest is exported, mainly to Japan and Europe.

Lying just within the Northwest Territories is the Canada Tungsten Mining Corporation's mine at Tungsten. Access is by nearly 200 miles (320 km) of road from the Alaska Highway and the mine is more a part of the economy of the Yukon Territory. It has been producing about 3.2 million lb (1,450 tonnes) of tungsten a year.

Two other mines are currently in operation in the Mackenzie District. Echo Bay Mine and Terra Mining and Exploration are both near Great Bear Lake and produce silver with a little copper at an annual rate of about 2½ million and 5 million oz (80 and 160 tonnes) respectively. Some other mines, such as the Rayrock Uranium Mine and the Discovery Gold Mine, were operated for a number of years but closed down when the ore was exhausted. A number of small silver properties have also been worked for a year or two, mining pockets of high-grade ore.

The mineral production of the Northwest Territories has remained at much the same level over the past ten years, though the dollar value has increased (see Table 5). Some 80 per cent of the value of production comes from one mine, Pine Point, which is on the southern edge of the Mackenzie District. Employment in the producing mines in 1975 totalled 1,583. Prospecting has indicated extensive mineralization in many areas, but current metal prices cannot compensate for the high costs of operating in the north, the short transport season which necessitates extensive inventories and stockpiling of product, and the lack of social infrastructure. At present economic factors do not appear to be favourable for expansion of the mining industry in the Northwest Territories.

In the Mackenzie District, as in the Yukon, only a small part of the hydroelectric potential has been developed. Most requirements are small and more easily met by diesel generators. To the 4 MW developed before the Second World War by the Consolidated Mining and Smelting Company near Yellowknife, the Northern Canada Power Commission has added installations on the Snare River 90 miles (140 km) north of Yellowknife which now produce 28 MW to serve the Yellowknife area, and on the Taltson River producing 22 MW for the area south of Great Slave Lake.

In the Keewatin District prospecting has shown extensive mineralization, but there has been only one producing mine. This was at Rankin Inlet where a high-grade nickel-copper ore was mined from 1957 to 1962 at a rate of about 250 tonnes a day. The mine closed when the ore was exhausted. The government encouraged Eskimos to seek employment at the mine, and a number of families moved there from other settlements. As Rankin Inlet is not in a good hunting area and had no economic base when the mine closed, it has been made into the administrative centre for the District.

The District of Franklin saw the first mine in Canada at the end of the sixteenth century when Frobisher believed he had found a gold ore in Frobisher Bay. It proved worthless and, apart from a few tons of mica from southern Baffin Island and a little coal for local consumption at Pond Inlet, no other minerals were mined until 1976, when a lead-zinc operation went into production at Strathcona Sound in northern Baffin Island. The Nanisivik mine is planned to mill 1,500 short tons (1,380 tonnes) of ore a day throughout the year and to ship the concentrates to Europe during the summer. In order to encourage the development, which would otherwise be at best marginal, and in this way to provide employment for Eskimos, the federal government has provided very considerable assistance in return for an 18 per cent equity interest.

A larger and richer lead-zinc discovery on Little Cornwallis Island will probably be developed by Cominco in the course of the next few years. Another large mineral discovery is exceptionally high-grade iron ore at Mary River in northern Baffin Island. The ore would be easy to mine and a railway could be built to the coast, but the shipping season would be short and development has been deferred.

Petroleum

During the years immediately following the Second World War there was little interest in the petroleum potential of the Canadian Arctic. Canadian attention was focused on Alberta, where the oil strike at Leduc near Edmonton in February 1947 was followed by other significant discoveries. Drilling gradually extended to the north of the provinces and over the border into the Northwest Territories. The enormous potential of the Athabasca oil sands had long been recognized and work continued on methods of releasing the oil from the sand. In 1967 Great Canadian Oil Sands Limited, controlled by Sun Oil Company, opened a plant now producing over 55,000 barrels of synthetic crude oil a day (3 million tonnes

Table 5. Mineral production in the Canadian northern Territories, 1966–76

Mineral		1966	1967	1968	1969	1970	1971	1972	1973	1974	1975	1976 (provisional)
Northwest Territories												
Gold	$	15,990,133	14,356,476	13,285,459	12,381,240	12,168,776	10,897,934	17,713,250	24,262,894	28,651,414	28,754,047	23,120,000
	oz	424,029	380,304	352,306	328,502	332,844	308,339	307,479	249,075	184,467	175,564	188,000
Silver	$	2,325,407	3,429,755	8,677,365	3,910,888	5,114,587	4,574,616	6,778,965	13,691,789	17,669,851	8,883,385	14,885,000
	oz	1,662,192	1,980,228	3,751,563	2,026,367	2,764,642	2,932,446	4,059,261	5,420,344	3,817,207	1,971,457	3,475,000
Copper	$	672,065	538,077	833,169	643,761	766,578	727,595	577,416	1,106,319	840,719	526,889	660,000
	lb	1,496,805	1,131,126	1,732,160	1,251,723	1,320,502	1,378,021	1,133,767	1,734,178	1,084,505	826,480	964,000
Nickel	$	—	—	—	—	—	—	—	—	—	—	—
	lb	—	—	—	—	—	—	—	—	—	—	—
Lead	$	31,472,562	35,665,535	33,636,984	32,299,014	37,842,405	32,629,795	27,838,277	32,261,787	34,932,761	37,254,292	26,808,000
	lb	210,659,720	254,753,820	250,275,180	212,913,740	239,206,099	167,628,110	180,439,960	199,887,160	168,708,403	183,844,711	118,340,000
Zinc	$	57,128,344	60,852,900	57,504,129	68,275,481	76,004,563	65,056,384	64,792,006	87,541,226	132,251,480	106,650,304	119,685,000
	lb	378,333,400	419,964,800	407,830,700	448,296,000	477,115,900	448,633,500	339,741,000	362,549,600	378,944,069	284,400,810	318,142,000
Cadmium	$	2,769,372	2,551,920	774,060	675,136	737,632	301,476	205,436	61,152	—	1,027	—
	lb	1,073,400	911,400	271,600	191,800	207,200	155,400	81,200	16,800	—	301	—
Bismuth	$	—	—	—	—	3,072	41,149	—	—	—	—	—
	lb	—	—	—	—	490	7,578	—	—	—	—	—
Tungsten[a]	$	—	—	—	—	—	—	3,174,120	3,228,600	3,557,600	3,257,840	5,100,000
	lb	—	—	—	—	—	3,288,400	—	—	—	—	—
TOTAL	$	110,357,883	117,394,663	114,711,166	118,185,520	132,637,613	114,228,949	117,905,350	158,925,167	214,346,225	182,069,944	185,158,000

Yukon Territory

Gold	$	1,639,103	675,725	911,338	1,118,715	653,034	511,534	234,983	2,032,502	4,111,631	5,255,077	3,910.000
	oz	43,466	17,900	24,167	29,682	17,862	14,473	4,079	20,865	26,472	32,086	31,000
Silver	$	5,868,217	6,701,756	4,806,384	5,182,166	7,845,312	8,966,417	8,331,575	15,342,856	26,800,905	28,531,397	13,446.000
	oz	4,194,580	3,869,374	2,077,987	2,685,060	4,240,709	5,747,703	4,988,967	6,073,973	5,789,783	6,331,868	3,139,000
Lead	$	2,386,684	2,141,959	970,629	4,256,183	20,830,196	29,340,379	34,392,366	38,013,324	41,194,600	54,888,680	19,104,000
	lb	15,975,125	15,299,709	7,221,940	28,056,581	131,670,010	217,336,142	222,921,742	235,522,452	198,950,056	270,867,945	84,335,000
Copper	$	—	3,409,779	5,097,157	7,645,623	9,148,995	2,709,696	890,286	14,791,665	15,571,426	11,928,559	16,639,000
	lb	—	7,167,919	10,597,000	14,866,077	15,760,000	5,132,000	1,748,093	23,186,245	20,086,720	18,711,172	24,336,000
Coal	$	46,390	15,791	—	—	—	—	—	—	—	—	—
	short tons	5,670	1,912	—	6,039	10,908	21,026	18,435	19,601	17,027	25,713	10.000
Zinc	$	1,729,027	1,373,151	748,206	5,035,385	24,845,216	39,003,342	45,241,287	61,167,027	60,899,995	95,400,540	42.898.000
	lb	11,450,510	9,476,545	5,306,429	33,062,280	155,964,948	233,134,144	237,225,560	253,321,575	174,498,553	254,401,441	114,029,000
Cadmium	$	306,336	265,997	147,716	239,965	261,528	114,654	82,759	45,718	17,331	15,423	12.000
	lb	118,735	94,999	51,830	68,172	73,463	59,100	32,711	12,560	4,358	4,519	5,000
Asbestos	$	—	406,371	8,684,125	11,924,526	13,927,652	12,374,380	13,006,476	13,915,140	22,752,400	32,820,720	34,460.000
	short tons	—	2,260	63,592	87,437	105,638	91,969	101,888	100,734	90,896	114,348	113,000
Nickel	$	—	—	—	—	—	—	3,006,762	5,209,621	—	—	—
	lb	—	—	—	—	—	—	2,814,621	3,404,981	—	—	—
Platinum	$	—	—	—	—	—	—	325,573	149,458	—	—	—
	oz	—	—	—	—	—	—	3,625	1,314	—	—	—
TOTAL	$	11,975,757	14,990,529	21,365,555	35,402,563	77,511,933	93,020,402	106,502,067	150,667,311	171,348,288	228,840,396	130,469,000

a Figures for value of tungsten production are not available.
(*Source*: Department of Indian Affairs and Northern Development, Ottawa.)

a year) from the sands. This is being followed by a larger plant now being built by Syncrude Canada, in which the Canadian, Alberta, and Ontario governments are participating with Imperial Oil, Gulf Oil Canada, and Canada-Cities Service, at an estimated cost of $2.2 billion. Production is planned to begin late in 1978 and to reach 125,000 barrels of synthetic crude a day (6 million tonnes a year) by 1981.

Farther north, considerable exploration at Norman Wells in connection with the wartime Canol project had shown that the oilfield there was modest in size, with recoverable reserves currently estimated at 50 million barrels (7 million tonnes). It has continued to supply the market along the Mackenzie River and the Western Arctic coast. Some exploratory drilling was carried out on the Peel Plateau and elsewhere in the Yukon and the Mackenzie Valley, but world supplies of oil appeared to be ample and the international oil companies could see no object in investigating the little-known Arctic, where exploration would be very costly and which was far away from possible markets. The petroleum possibilities of the Canadian Arctic did not begin to be recognized until reconnaissances by the Geological Survey of Canada in the early 1950s had indicated the extent and thickness of the sedimentary strata there. In 1959 the Canadian government drew up new regulations governing petroleum exploration in the far north. They were intended to encourage the oil industry, and their publication was followed by a flood of applications covering most of the sedimentary areas in the Arctic islands and parts of the lower Mackenzie Valley.

In 1962 the first well in the Arctic islands was completed at Winter Harbour in Melville Island and it was followed in 1963 by drilling at Resolute Bay on Cornwallis Island and the next year on Bathurst Island. Owing to the expense of these operations and their lack of success, interest in the Arctic islands waned. The major oil companies preferred to spend their exploration funds in the North Sea, where exciting discoveries had been made, and in the lower Mackenzie Valley, which was closer to markets and where there were fewer uncertainties. They felt they could afford to wait indefinitely before becoming involved in the Arctic islands. To break this impasse the Canadian government encouraged in 1967 the establishment of Panarctic Oils, a consortium of exploration companies who pooled their acreage in the far north and, with 45 per cent equity. participation by the Canadian government, subscribed to finance an extensive program of exploration.

Before Panarctic had drilled its first well, the discovery of the largest oilfield in North America at Prudhoe Bay on the Arctic coast of Alaska provided a tremendous impetus to oil exploration in the Canadian north. Contrary to the practice in many countries, the whole of the Canadian north was available for exploration, with no part held back for later disposal. Petroleum exploration permits were sought and issued for the remaining sedimentary land areas and far out on the continental shelf. Exploration increased rapidly and results were at first encouraging, substantial amounts of gas being discovered in both the Sverdrup Basin and the Mackenzie Delta. Exploratory drilling increased from 40,000 metres in 1968 to reach a peak of nearly 200,000 metres in 1973 when eighty-three holes were drilled in the Territories. The later drilling proved disappointing. Over $500 million has been spent on exploration in the Arctic islands, but little oil has been found and the gas reserves have not increased as fast as had been anticipated. In the Mackenzie Delta area also, initial successes raised expectations that further drilling failed to sustain. Since 1973 petroleum exploration has declined throughout the north, despite the growing concern regarding national fuel suplies; the number of wells drilled in 1975 had fallen to forty-two and total exploration expenditures from $263 million to $215 million. Interest is now centred on the Canadian sector of the Beaufort Sea, where geological conditions appear to be very favourable, which is in the same general area as Prudhoe Bay, and which is only now being tested by drilling. The costs of offshore exploration in these waters are, however, very high, largely because of the short drilling season and the environmental safeguards required by the government.

Differing opinions are held on the potential reserves of gas and oil in the Canadian north, and radical revisions of the figures announced by the oil companies have discounted their accuracy. The most recent (1976) estimates of the Department of Energy, Mines, and Resources for the Mackenzie Delta–Beaufort Sea area are 6.5 trillion cu ft (180 billion cu m) of gas and 400 million barrels (55 million tonnes) of oil proved, with a 90 per cent probability of 39 trillion cu ft (1.1 trillion cu m) of gas and 4,300 million barrels (580 million tonnes) of oil (see Tables 6 a and b). For the Sverdrup Basin and the folded belts immediately to the south of it the figures are 10 trillion cu ft (280 billion cu m) of gas proved, with a 90 per cent probability of 23.9 trillion cu ft (680 billion cu m), and for oil 100 million barrels (14 million tonnes) proved with a 90 per cent probability of 1,600 million barrels (220 million tonnes). Reserves are stated in the form of graphs, of which Table 6(b) is an example. Table 6(a) summarizes this information for a number of areas of northern Canada.

Oil and gas production in the Territories now comes from three fields. Norman Wells is the only oil producer with an output of about 1 million barrels (140,000 tonnes) a year. Beaver River and Pointed Mountain are gas-producing fields. The Beaver River field lies mainly in British Columbia but extends into the Yukon. Production began in 1971 and reached 250 million cu ft (7 million cu m) a day but has declined owing to water intrusion into the reservoir. The nearby Pointed Mountain field

Table 6a. Petroleum reserves in northern Canada

Regions	Probability (per cent)											
	100	90	80	70	60	50	40	30	20	10	5	0
Gas Reserves[a]												
Northern mainland[b]	1.0	6.0	7.0	7.9	8.8	9.7	11	12	16	20	26	64
Mackenzie Delta–Beaufort Sea[c]	6.5	39	45	50	55	60	65	72	81	99	121	339
Sverdrup Basin[d]	10	21	27	31	36	40	46	53	63	80	101	169
Arctic fold belts[e]	0	2.9	5.7	8.0	9.4	11	12	14	16	26	50	150
Labrador–East Newfoundland shelf[f]	8	18	20	23	25	27	29	32	36	45	55	242
Oil Reserves[h]												
Northern mainland	0.05	0.34	0.39	0.44	0.49	0.54	0.61	0.69	0.81	1.0	1.3	3.1
Mackenzie Delta–Beaufort Sea	0.40	4.3	5.1	5.7	6.3	6.9	7.6	8.5	9.7	12	16	36
Sverdrup Basin	0	1.1	1.3	1.6	1.8	2.0	2.3	2.6	3.1	4.0	5.1	8.5
Arctic fold belts	0.1	0.50	0.95	1.3	1.6	1.8	2.0	2.3	2.7	4.3	8.4	25
Labrador–East Newfoundland shelf	0	1.7	2.0	2.2	2.4	2.6	2.9	3.2	3.6	4.5	5.7	34

Some areas are not shown in the table because their oil and gas possibilities are poor, e.g. Hudson Bay, Victoria Island, or because they are beyond current drilling capabilities or are under ice, e.g. the Baffin Bay shelf.
[a] Gas reserves are in trillions of cubic feet (1 trillion cu ft = 28 billion cu m)
[b] The Northern mainland includes the continuation of the Western Canadian Sedimentary Basin north from 60° N to the Mackenzie Delta but excludes the Delta.
[c] The Mackenzie Delta–Beaufort Sea region extends to the edge of the continental shelf.
[d] The Sverdrup Basin lies in the northern part of the Queen Elizabeth Islands.
[e] The Arctic fold belts form a band south of the Sverdrup Basin.
[f] The Labrador–East Newfoundland shelf extends to water depths of up to 1500 ft (460 m).
[g] Gas discoveries have been made on the Labrador shelf but reliable estimates of these reserves are not yet available.
[h] Oil reserves are in billions of barrels (1 billion barrels = 140 million tonnes).
(*Source*: *Oil and Natural Gas Reserves of Canada 1973*, Department of Energy, Mines, and Resources, Ottawa, Report EP 77–1.)

Table 6b.

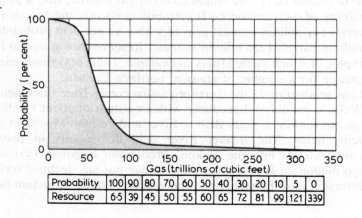

Probability	100	90	80	70	60	50	40	30	20	10	5	0
Resource	6.5	39	45	50	55	60	65	72	81	99	121	339

lies wholly within the Northwest Territories and produces up to 100 million cu ft (2.8 million cu m) a day. A spur from the Westcoast Pipeline takes the gas from both fields to the south via Fort Nelson.

Other resources

In both the Yukon and Northwest Territories, *tourism* has been encouraged, despite the short season and the consequent limited utilization of accommodation and other tourist facilities. The Yukon Territory has the advantages of a longer season, a much more developed road system, the historic interest of the Klondike gold rush, and excellent mountaineering. It is also more accessible, especially from Alaska from which it attracts many visitors. In 1975 the number of visitors by road, air, or rail was over 340,000, and their estimated direct expenditures were over $27 million.

Tourism has grown much more slowly in the more remote Northwest Territories. A journey there by road entails a long and dusty drive, air travel is expensive, and it is difficult to reach the most spectacular scenery. Fishing lodges in remote areas attract a number of rich anglers, and package tours, arranged to visit the more interesting places or catering to some special interest, are becoming increasingly popular. In 1975 the number of tourists that went to the Northwest Territories was 21,000, and their total expenditures were estimated to be over $10 million.

The rate of growth of tourism in the north will depend to a large extent on the roads. At present there are few places in the Northwest Territories to which a tourist can drive, the roads are gravel-surfaced, and repair services and other facilities are far apart. Tourism is also an industry that is sensitive to changes in general prosperity. In the long term, increasing numbers of visitors will be attracted to the northern National Parks, of which four have now been established and others are being planned.

In the provinces *forestry* operations have continued to move farther north with the encouragement, and sometimes the financial participation, of provincial governments. Consideration is being given in Ontario to allowing cutting in the northwest in what are almost the last remaining stands of untouched bush in the province. There is growing opposition owing to the disturbance to the life of the Indians who live there, the very slow rate of regeneration, and fear of pollution of the rivers. The western provinces have seen the construction since the Second World War of a number of very large mills as at The Pas, Prince Albert, Hinton, Grande Prairie, Prince George, Mackenzie, and Port Edward. The forestry industry has been a very powerful force in opening up the northland to the economy of southern Canada. The scale of modern operations is so large that, as with the mining industry, towns can be created almost overnight, and attract secondary and service industries to sustain their growth. Unlike mining communities they are based on a renewable resource, and do not have the same fear of becoming ghost towns.

There are substantial forested areas in both the Yukon Territory and the Mackenzie District, particularly in the river valleys and in the south, and they supply lumber, fuel, and mining timbers for local use. Productive forest covers 42,100 sq miles (110,000 sq km) in the Yukon Territory and 33,600 sq miles (85,000 sq km) in the Northwest Territories, with an estimated 9,000 million and 14,000 million cu ft (250 and 400 million cu m) of merchantable standing timber respectively. Most of this is coniferous.

Production from the northern forests could be expanded many times over if markets could be developed. Even in the north, however, lumber has difficulty in competing with supplies from outside the territories, and oil is a cheaper and much more convenient source of fuel. There has been some increase in the annual cut recently. In the Yukon Territory it now exceeds 3 million cu ft (85,000 cu m), mainly from the Watson Lake area, against an average over the past twenty-five years of 2 million cu ft (55,000 cu m). In the Northwest Territories most forestry operations are south and west of Great Slave Lake and annual production has also increased, reaching nearly 2 million cu ft against a former average of about 1 million (28,000 cu m).

With recent increases in prices, the annual value of forest products from both Territories has been approaching $5 million. This is considerably less than the annual expenditures on forest protection, where fire-fighting costs are nearly $6 million, most of which is spent on aircraft charter and wages for fire-fighting crews.

The catch of *fish* in 1975 in the Northwest Territories was 2,568,000 lb (1,160 tonnes) with a landed value of $677,000. There has been a continuous decline in the industry for several years, and the catch in 1976 was little more than half of the 1970 total. The Great Slave Lake fishery is much the largest, and whitefish the most important species. Farther north Arctic char is caught by the Eskimos both for their own use and for export to southern Canada where there is a ready market. The chief constraints on expansion of char fishing are the difficulty and expense of transporting fish from the north without sacrificing quality, and the low productivity of cold northern waters which necessitates the setting of low quotas.

Favoured areas of the Yukon, the Mackenzie District, and Labrador have some limited *agriculture* potential, and in the first half of the present century a few head of cattle were kept and vegetables were grown in very limited quantities. Every improvement in transport to the north has, however, made it more difficult for agriculture in the north to succeed, as it has eroded the protection provided by isolation. Fresh food can now be flown or trucked in from the south to the areas where agriculture was struggling to provide what had been unobtainable.

Fur continues to make a useful contribution to the northern economy, and the value of fur produced has in fact risen in recent years to $2,750,000 in the Northwest Territories in 1976 and $400,000 in the Yukon Territory in 1975. North of the treeline sealskins have been added to white-fox pelts as a valuable resource. In some areas, such as Banks Island, trapping in good years at least compares favourably with wage employment as a source of income. Most settlements in the north still rely on hunting and fishing for much of their food.

The part that *hunting* plays in the local economy is often understated or ignored in statistical reports because it rarely results in cash transactions. In the smaller northern settlements the value of the food obtained through hunting is probably equivalent to an annual income of about $4,000 for each family. Hunting and trapping have, however, more than just an economic importance. They are associated with the traditional Eskimo life and even in areas where their contribution to income is small they represent much more than a recreation.

A resource possibly of very great potential value lies in the abundant *fresh water* of the north. Consumption of water in the developed regions of North America is increasing, particularly in parts of the United States, and proposals have been made to divert northern rivers to flow to the south. Undertakings of this sort would have great political implications and would be extremely costly, and they lie far in the future.

An important supplement to the economy of the Northwest Territories has been the production of Eskimo *graphics* and *soapstone carvings* which have found a ready market in both America and Europe. Fur garments and other handicrafts are also produced by the Eskimos for sale in the south. These activities have been strongly encouraged by the government. Much of the marketing for the Northwest Territories is handled centrally through Canadian Arctic Producers, a company originally established by the government and now in the process of becoming an Eskimo co-operative. The return to the natives of the Northwest Territories from this work is in the order of $2.5 million a year, most of it from carvings and prints.

Eskimo graphics and carvings are produced also in New Quebec, with much of the marketing in the hands of La Fédération des Coopératives du Nouveau Québec, and yield a return of about $1.8 million a year.

Population

All elements of the population of both northern Territories are increasing, the native population by natural increase and the white mainly by immigration (see Table 7). The compositions of the populations are very different. The native population is much younger, with a median age of just under 16 years and the sexes are nearly in balance, with the female to male ratio of .983. The white population has a median age of almost

Table 7. Population of the Canadian northern Territories, 1941–76

	Northwest Territories					Yukon Territory					
Year	Indian	Eskimo	Métis[a]	Other	Total	Indian	Eskimo	Métis[a]	Other	Total	TOTAL
1941	9,456	282	2,290		12,028	1,508		193	3,213	4,914	16,942
	78.6%		21.4%			30.7%		69.3%			
1951	10,660		5,344		16,004	1,563			7,533	9,096	25,100
	66.6%		33.4%			17.2%		82.8%			
1961	13,233		9,765		22,998	2,207			12,421	14,628	37,626
	57.5%		42.5%			15.1%		84.9%			
1971	7,180	11,400		16,225	34,805	2,580	10		15,800	18,390	53,195
	20.6%	32.9%	46.6%			14.1%		85.9%			
1976	8,450	13,000	2,800	17,990	42,240	3,240	10	1,200	16,945	21,395	63,635
	20.6%	32.6%	46.8%			15.2%		84.8%			

[a] The figures for Métis, when shown, are rough approximations, Métis being those whose male parent was white and female parent was native. Those whose male parent was native and female parent was white are included as native.
(*Sources: Statistics Canada: Decennial censuses*, for 1941, 1951, 1961 and 1971, Ottawa. *Statistics Canada: 1976 Census – population: preliminary counts*, Ottawa.)

22½ years and there are many more men than women, the female to male ratio being .827.

For several years the government has been encouraging family planning and this is reflected in the birth-rate. In the Northwest Territories it had fallen from 41.7 live births per 1,000 population in 1968 to 27.8 in 1975, but it rose to 32.4 the next year. This is about twice the national average. It is particularly high among the Eskimos, but has dropped substantially from nearly 50 in 1969.

Deaths from accidents, injuries, and violence have increased in the past ten years, from about 20 per cent of all deaths in the Northwest Territories to 30 per cent. Together they are responsible for over twice as many deaths as any other major cause, such as heart disease or pneumonia.

In the Yukon Territory Indians form 15 per cent of the population, but only 700 individuals live in Whitehorse, which has over 12,000 whites, and 61 per cent of the total population of the Territory. Elsewhere in the Yukon the Indians form about one-third of the population. There are fewer than a dozen Eskimos, who live on the Arctic coast of the Yukon.

In the Northwest Territories the white population is concentrated in the Mackenzie Valley, especially in the larger towns of Yellowknife, Hay River, Fort Smith, Fort Simpson, and Inuvik. The Eskimos are mainly in the Keewatin and Franklin Districts, where they formed 82.8 per cent of the population in the 1971 census. The white population of these Districts totals 1,880, most of whom live in Frobisher Bay or are engaged in activities related to petroleum exploration in the Queen Elizabeth Islands.

Transport

Roads

Winter roads have a long history in the north. A winter road was built from Norway House to York Factory in 1825, and they were widely used in the Yukon at the turn of the century. The development of a year-round road network had its origins in the Second World War. The Alaska Highway provided road access to the Yukon from the south and became the trunk from which the road system of the Territory has branched out. By the end of the war the Haines Road connected the Highway to the Pacific coast and the Canol Road crossed the Mackenzie Mountains to Norman Wells. The old winter roads to Atlin, Mayo, and Dawson City were improved for year-round use and it was the central position of Whitehorse in the transport complex of road, rail, air, and river, and its consequent population growth that led to the move of the Territorial capital there from Dawson City in 1953. Road transport is now vital to the economy of the Yukon. New roads like the Campbell Highway from Watson Lake to Carmacks via Ross River have played a very important part in mineral exploration, and it is by road that the outputs of the major mines in the Territory, at Clinton Creek, Keno Hill, Faro, and in the Whitehorse area, as well as the asbestos fibre from Cassiar in northern British Columbia are trucked to Whitehorse.

In 1958 the government began construction of a road north from Dawson to Chapman Lake and it was later decided to extend it to Fort McPherson and Arctic Red River in the Mackenzie Delta. This road, the Dempster Highway, was nearing completion in 1977 when it was passable except for the last 36 miles (58 km) in the Yukon before reaching the border of the Northwest Territories. As Arctic Red River is connected by road to Inuvik, it will be possible to travel by road from southern Canada to Inuvik. Several other all-weather roads have been built in the Yukon Territory and another important road, now under construction and running from Carmacks to Skagway in Alaska, will be completed in 1978. The section of the Canol Road in the Yukon has been maintained, but not the part in the Northwest Territories.

In the Mackenzie District the first all-weather road was the Mackenzie Highway from Grande Prairie to Hay River. Construction began in 1949 and in the 1950s the Yellowknife Highway was built to lead from it around

Great Slave Lake to Fort Providence, Rae, and Yellowknife. Roads were also built to Fort Smith, Pine Point, and Fort Resolution. The Mackenzie Highway was later extended to Fort Simpson, and in 1972 the government decided to continue it all the way down the river to Inuvik and possibly Tuktoyaktuk. This met with strong resistance from many of the Indians, who recognized the proposed road as a serious threat to their way of life. At the same time uncertainty about the Mackenzie Valley pipeline and a quadrupling of cost estimates gave the government second thoughts about the priority of a road along the Mackenzie River, which is already well served by barge traffic in the summer, and in 1975 it was decided to terminate construction for the time being at least at Wrigley (see Map 11).

There are no roads in the Keewatin and Franklin Districts of the Northwest Territories, but the James Bay hydroelectric development has led to the construction of a road from Mattagami to Fort George. This provides the first and only access to Hudson Bay by road.

Railways
The years after the Second World War saw the construction of several railways in the northland. Most were built to serve a single mine, often as spurs from a main line. They were not planned to meet regional needs and did not have much effect on other development. An exception was the Pine Point Railway, which was built primarily to transport lead and zinc concentrates from Pine Point and was partly financed by the mining company. The line runs also to Hay River and it has become an important element in the supply system along the Mackenzie Valley. The Pine Point Mine now accounts for only half the traffic carried. An extension of the line all the way to the Arctic coast has been suggested as a way of transporting petroleum to the south, but it would be more expensive to construct than a pipeline and could have very heavy maintenance costs.

The Yukon Territory is still without a railway link to the south. Studies have been made of the feasibility of building a line north from the British Columbia Railways to the Yukon, and of extending the White Pass and Yukon Railway from Whitehorse to Carmacks, but neither line is likely to be built for several years.

Water
After the Second World War transport along the Mackenzie River greatly increased. The Northern Transportation Company had been taken over by the Canadian government along with its uranium-producing parent, Eldorado Mining and Refining Limited, and it became a crown corporation. It absorbed the Mackenzie River Transport Division of the Hudson's Bay

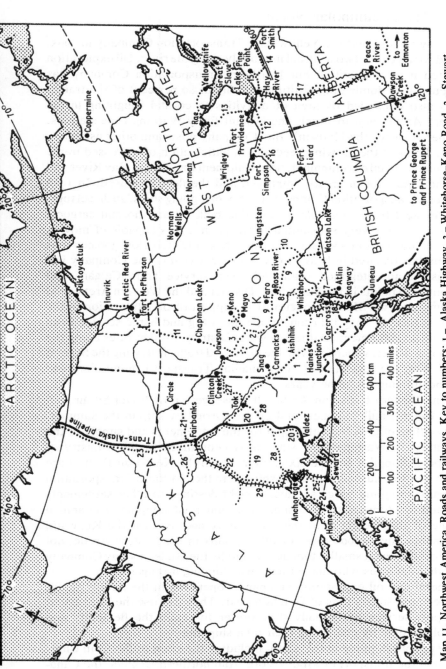

Map 11. Northwest America. Roads and railways. Key to numbers: 1 – Alaska Highway, 2 – Whitehorse–Keno Road, 3 – Stewart Crossing–Dawson Road, 4 – Haines Road, 5 – Carcross–Skagway Road, 6 – Tagish Road, 7 – Atlin Road, 8 – Canol Road, 9 – Campbell Highway, 10 – Nahanni Range Road, 11 – Dempster Highway, 12 – Mackenzie Highway, 13 – Yellowknife Highway, 14 – Fort Smith Highway, 15 – Pine Point–Ft. Resolution Highway, 16 – Liard Highway, 17 – Pine Point Railway, 18 – White Pass and Yukon Railway, 19 – Denali Highway, 20 – Richardson Highway, 21 – Steese Highway, 22 – Anchorage–Fairbanks Highway, 23 – Pipeline Road, 24 – Kenai Highway, 25 – Seward Highway, 26 – Elliott Highway, 27 – Taylor Highway, 28 – Glenn Highway, 29 – Alaska Railroad.

Company in 1947 and the Yellowknife Transportation Company in 1965, to hold for a year or two a virtual monopoly along the river. Oil exploration attracted new competition but Northern Transportation Company has continued in a dominating position with about 80 per cent of the traffic. It has also expanded its area of activities to coastal freighting in the Western Arctic, first to supply the Distant Early Warning Line stations, and then to serve the Hudson's Bay Company posts and other establishments. To meet the growing need and in anticipation of increased traffic associated with oil development, trans-shipment facilities at Hay River and elsewhere have been greatly improved and the fleet increased. In 1976 Northern Transportation Company had 3 ocean-going ships, 29 diesel tugs, and 167 steel barges capable of carrying bulk oil and normal cargoes, which represents very substantial overcapacity. It is capable of moving about 560,000 tons in a season, but traffic has fallen to below 300,000 tons and this has resulted in heavy losses in recent years. The Company has established an integrated road and river service from Calgary and Edmonton to points in the Mackenzie District and the Western Arctic, and also employs two SRN-6 Hovercraft in support of oil exploration in the Beaufort Sea. In 1975 it began to operate a tug and barge service from Churchill to settlements in the Keewatin District. Freight rates along the Mackenzie River, which fell with the increased traffic following the Second World War, are regulated by the Canadian Transport Commission, and have been rising for some years.

In the Eastern Canadian Arctic all passengers now travel by air, but heavy freight, bulk supplies, and oil are carried by ship in the summer. The handling of government freight for the settlements and supply of the Distant Early Warning Line stations is centralized in the Ministry of Transport. Supplies for the six settlements in the Keewatin District are sent by rail to Churchill and trans-shipped to the Northern Transportation Company's tug and barge service. Freight destined to other settlements is routed to Montreal to be loaded on commercial shipping. The annual volume of this freight is in the order of 20,000 tons for the Keewatin settlements, and 60,000 tons for other points in the Eastern Arctic, not including the much smaller tonnage sent in by the Hudson's Bay Company and other agencies who make their own shipping arrangements.

Mineral and oil exploration companies operating in the Arctic islands require some 50,000 tons of freight a year. They can use the Ministry of Transport service but usually they prefer to retain the flexibility and control possible by chartering their own shipping.

To support shipping in the north, including the grain traffic from Churchill, and the transport of asbestos ore from northern Ungava, the Canadian Coast Guard has an icebreaking fleet. Six of these ships are heavy icebreakers (*Louis S. St Laurent, John A. MacDonald, d'Iberville, Lab-*

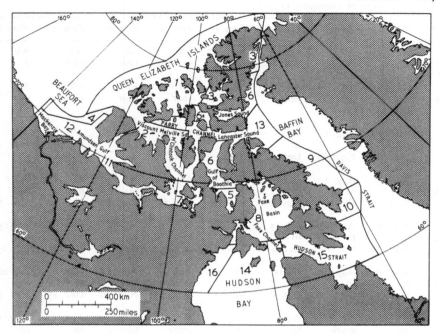

Map 12. Canada. Marine zones established under provisions of the Arctic Waters Pollution Prevention Act, 1971 (*Source:* Ministry of Transport, Ottawa)

rador, N. B. McLean, and *Norman McLeod Rogers*) and six are medium icebreakers (*Montcalm, J. E. Bernier, Sir Humphrey Gilbert, Sir William Alexander, Wolfe,* and *Camsell*). The *Camsell* operates in the Western Arctic and is based at Victoria. The other icebreakers are based on the Atlantic coast and are used in the St Lawrence during the winter. Construction of two new heavy icebreakers (Arctic Class 3) is proceeding, and design studies for icebreakers of higher classes are being made. In the classification system used, the number of the class gives the number of feet of ice through which an icebreaker of that class can proceed at a steady rate. Class 1A is intermediate between Class 1 and Class 2. None of the present Canadian icebreaking fleet exceeds Arctic Class 4.

There have been two recent developments of particular importance in relation to shipping in the Canadian Arctic. One is the proclamation of the Arctic Waters Pollution Prevention Act, 1971, designed *inter alia* to reduce the possibility of oil spills and other environmental damage in the Canadian north owing to the use of unsuitable ships. To assist in administering the Act, Canadian waters north of 60° N have been divided into sixteen zones based on the severity of their ice conditions (Map 12), and a schedule has been drawn up showing the periods of the year when

Table 8. Navigation periods permissible under the Arctic Waters Pollution Prevention Act 1971 (*Source*: Ministry of Transport, Ottawa)

Category	Zone 1	Zone 2	Zone 3	Zone 4	Zone 5	Zone 6	Zone 7	Zone 8
Arctic Class 10	All year	All year	All year	All year	All year	All year	All year	All year
Arctic Class 8	Jul. 1 to Oct. 15	All year	All year	All year	All year	All year	All year	All year
Arctic Class 7	Aug. 1 to Sept. 30	Aug. 1 to Nov. 30	Jul. 1 to Dec. 31	Jul. 1 to Dec. 15	Jul. 1 to Dec. 15	All year	All year	All year
Arctic Class 6	Aug. 15 to Sept. 15	Aug. 1 to Oct. 31	Jul. 15 to Nov. 30	Jul. 15 to Nov. 30	Aug. 1 to Oct. 15	Jul. 15 to Feb. 28	Jul. 1 to Mar. 31	Jul. 1 to Mar. 31
Arctic Class 4	Aug. 15 to Sept. 15	Aug. 15 to Oct. 15	Jul. 15 to Oct. 31	Jul. 15 to Nov. 15	Aug. 15 to Sept. 30	Jul. 20 to Dec. 31	Jul. 15 to Jan. 15	Jul. 15 to Jan. 15
Arctic Class 3	Aug. 20 to Sept. 15	Aug. 20 to Sept. 30	Jul 25 to Oct. 15	Jul. 20 to Nov. 5	Aug. 20 to Sept. 25	Aug. 1 to Nov. 20	Jul. 20 to Dec. 15	Jul. 20 to Dec. 31
Arctic Class 2	No entry	No entry	Aug. 15 to Sept. 30	Aug. 1 to Oct. 31	No entry	Aug. 15 to Nov. 20	Aug. 1 to Nov. 20	Aug. 1 to Nov. 30
Arctic Class 1A	No entry	No entry	Aug. 20 to Sept. 15	Aug. 20 to Sept. 30	No entry	Aug. 25 to Oct. 31	Aug. 10 to Nov. 5	Aug. 10 to Nov. 20
Arctic Class 1	No entry	No entry	No entry	No entry	No entry	Aug. 25 to Sept. 30	Aug. 10 to Oct. 15	Aug. 10 to Oct. 31
Type A	No entry	No entry	Aug. 20 to Sept. 10	Aug. 20 to Sept. 20	No entry	Aug. 15 to Oct. 15	Aug. 1 to Oct. 25	Aug. 1 to Nov. 10
Type B	No entry	No entry	Aug. 20 to Sept. 5	Aug. 20 to Sept. 15	No entry	Aug. 25 to Sept. 30	Aug. 10 to Oct. 15	Aug. 10 to Oct. 31
Type C	No entry	No entry	No entry	No entry	No entry	Aug. 25 to Sept. 25	Aug. 10 to Oct. 10	Aug. 10 to Oct. 25
Type D	No entry	No entry	No entry	No entry	No entry	No entry	Aug. 10 to Oct. 5	Aug. 15 to Oct. 20
Type E	No entry	No entry	No entry	No entry	No entry	No entry	Aug. 10 to Sept. 30	Aug. 20 to Oct. 20

Category	Zone 9	Zone 10	Zone 11	Zone 12	Zone 13	Zone 14	Zone 15	Zone 16
Arctic Class 10	All year	All year	All year	All year	All year	All year	All year	All year
Arctic Class 8	All year	All year	All year	All year	All year	All year	All year	All year
Arctic Class 7	All year	All year	All year	All year	All year	All year	All year	All year
Arctic Class 6	All year	All year	Jul. 1 to Mar. 31	All year	All year	All year	All year	All year
Arctic Class 4	Jul. 10 to Mar. 31	Jul. 10 to Feb. 28	Jul. 5 to Jan. 15	June 1 to Jan. 31	June 1 to Feb. 15	June 15 to Feb. 15	June 15 to Mar. 15	June 1 to Feb. 15
Arctic Class 3	Jul. 20 to Jan. 20	Jul. 15 to Jan. 25	Jul. 5 to Dec. 15	June 10 to Dec. 31	June 10 to Dec. 31	June 20 to Jan. 10	June 20 to Jan. 31	June 5 to Jan. 10
Arctic Class 2	Aug. 1 to Dec. 20	Jul. 25 to Dec. 20	Jul. 10 to Nov. 20	June 15 to Dec. 5	June 25 to Nov. 15	June 25 to Dec. 10	June 25 to Dec. 20	June 10 to Dec. 10
Arctic Class 1A	Aug. 10 to Dec. 10	Aug. 1 to Dec. 10	Jul. 15 to Nov. 10	Jul. 1 to Nov. 10	Jul. 15 to Oct. 31	Jul. 1 to Nov. 30	Jul. 1 to Dec. 10	June 20 to Nov. 30
Arctic Class 1	Aug. 10 to Oct. 31	Aug. 1 to Oct. 31	Jul. 15 to Oct. 20	Jul. 1 to Oct. 31	Jul. 15 to Oct. 15	Jul. 1 to Nov. 30	Jul. 1 to Nov. 30	June 20 to Nov. 15
Type A	Aug. 1 to Nov. 20	Jul. 25 to Nov. 20	Jul. 10 to Oct. 31	June 15 to Nov. 10	June 25 to Oct. 22	June 25 to Nov. 30	June 25 to Dec. 5	June 20 to Nov. 20
Type B	Aug. 10 to Oct. 31	Aug. 1 to Oct. 31	Jul. 15 to Oct. 20	Jul. 1 to Oct. 25	Jul. 15 to Oct. 15	Jul. 1 to Nov. 30	Jul. 1 to Nov. 30	June 20 to Nov. 10
Type C	Aug. 10 to Oct. 25	Aug. 1 to Oct. 25	Jul. 15 to Oct. 15	Jul. 1 to Oct. 25	Jul. 15 to Oct. 10	Jul. 1 to Nov. 25	Jul. 1 to Nov. 25	June 25 to Nov. 10
Type D	Aug. 15 to Oct. 20	Aug. 5 to Oct. 20	Jul. 15 to Oct. 10	Jul. 1 to Oct. 20	Jul. 30 to Sept. 30	Jul. 10 to Nov. 10	Jul. 5 to Nov. 10	Jul. 1 to Oct. 31
Type E	Aug. 20 to Oct. 15	Aug. 10 to Oct. 20	Jul. 15 to Sept. 30	Jul. 1 to Oct. 20	Aug. 15 to Sept. 20	Jul. 20 to Oct. 31	Jul. 20 to Nov. 5	Jul. 1 to Oct. 31

different categories of icebreaker and ice-strengthened ships are permitted to operate (Table 8). Shipping that does not meet the necessary requirements can be ordered to leave the zone, but compliance with unilateral legislation of this nature is difficult to enforce.

The other important development is the ordering by a consortium of Canadian shipping companies of the *Arctic*, an icebreaking cargo vessel of 28,000 dwt and Arctic Class 2 category. She is due for completion in 1978 and much of the cost is being borne by the Canadian government. She is the first Canadian cargo vessel to have been designed as an icebreaker as distinct from a ship strengthened for use in ice, and it is planned to use her initially to carry lead-zinc concentrates from Strathcona Sound to Germany from late June to mid-November each year. If she proves capable of extending the shipping season appreciably, the development of other mineral properties in the Arctic could become feasible.

A different approach to the problem of shipping in the north is being tested with an icebreaking Arctic Class 3 barge of 11,000 dwt, powered by two-9000 hp conventional tugs fitting into two deep notches at the stern. Experiments are also being carried out to determine the icebreaking capacity of air-cushioned vehicles, which gives promise of providing a new and effective technique.

Petroleum

Recognition of the possibility of finding petroleum in the Canadian north brought with it the question of how any petroleum that was discovered could be transported to markets in the south. Related issues, in particular the protection of the environment, native land claims, declining petroleum reserves in Canada and the United States, and increasing world petroleum prices have often obscured the actual transportation problems.

Significant quantities of hydrocarbons have been found in two regions of the Canadian North – the Mackenzie Delta area and the Sverdrup Basin of the Canadian Arctic Archipelago. Sufficient exploration has been carried out to establish that, while further drilling will increase the reserves considerably, very major accumulations of petroleum are unlikely to be found in the north except possibly off-shore in the Beaufort Sea or Davis Strait.

In the Mackenzie Delta about 6.5 trillion cu ft (180 billion cu m) of gas have been found. Oil has not yet been proved in commercial quantities. The only practical way to move the gas appears to be by pipeline. Reserves of at least 15 trillion cu ft (400 billion cu m) would be needed to justify the cost of a pipeline up the Mackenzie Valley. Very large quantities of gas have however been found in the Prudhoe Bay field in Alaska and this will have to be piped to the south. The oil companies had hoped to route the pipeline from Prudhoe Bay through the Mackenzie Delta so that it could also carry gas from there. Both the Berger Commission and the Canadian National Energy Board, however, rejected this plan on environmental grounds, and the United States and Canadian governments subsequently agreed that the Prudhoe Bay gas would be carried through northern Canada by a pipeline to be built along the route of the Alaska Highway. The Mackenzie Delta and any Beaufort Sea gas will eventually, in all probability, be carried by a pipeline running south, possibly for much of the way along the Dempster Highway, to join the Alaska Highway pipeline in the Yukon Territory. This line would not require such large reserves of gas to justify its cost. Timing of construction will depend on many interests and issues: the oil companies anxious to retrieve their investment in exploration; the native people, worried about their own future and wanting to share, possibly through land claims, in the wealth that would be taken from the land where they have lived so long; the demand for gas and the rate at which additional reserves are found in both the north and southern Canada.

In the Sverdrup Basin a significant oil field, with 100 million barrels (14 million tonnes) now proved, has been located on Cameron Island. The field will probably be developed when reserves in the area have reached 500 million barrels (70 million tonnes) and the oil will be transported by ship. Three very large gas fields have been found, two on the Sabine Peninsula area of Melville Island, and the other south of Ellef Ringnes Island. Their combined proved reserves now total over 10 trillion cubic feet (280 billion cu m). The threshold that is needed to justify a pipeline is between 25 and 30 trillion cubic feet (700–850 billion cu m). There are, however, great technical difficulties in building a pipeline as it would have to cross several deep channels, where scouring by ice could be a hazard. An appealing solution lies in transporting gas in the form of liquefied natural gas by ship. Substantial energy losses are involved in the process of liquefaction, but it has the advantage of greater flexibility as the destination of the gas can be changed.

Liquefaction would also allow more flexibility in the rate of exploitation. In order to be economic a pipeline would have to be of large diameter, and would transport more gas than the Canadian market could easily handle. Liquefied gas would not require as high a threshold of reserves and could be extracted at a rate to suit Canadian needs, a point of considerable political importance in a country where heating is a necessity for survival and fuel supplies are limited. In addition, transportation by ship would have fewer environmental hazards and little direct impact on the life of the native people. A consortium of Petrocan, the recently established Canadian government oil company, and several leading Canadian shipping companies, is planning to carry out a pilot scale movement of liquefied natural gas from Melville Island to southern Canada in order to determine the feasibility of this method and its likely cost.

The capital requirements for moving gas by either pipeline or tanker from the Sverdrup Basin, and for building a pipeline from the Mackenzie Delta would be so large that one would compete with the other. Financing will therefore be an important factor in determining the timing of measures for moving northern petroleum.

Air

The pattern of air transport in the Canadian north has been set largely by the Air Transport Committee of the Canadian Transport Commission, which issues licences to the air carriers. In contrast to the situation in many countries, the Canadian government does not follow a practice of providing direct subsidies to air carriers. The rulings of the Air Transport Committee are, however, designed to encourage the carriers by making it possible for them to operate profitably. The Committee does not, therefore, issue licences to more carriers in any area or for any route than it considers the volume of traffic can support, and has the power to limit the use of airfields for charter operations to designated carriers. While such measures certainly benefit the established carriers, the lack of competitive services can lead to inefficiency and poor public relations.

The air-transport system that has developed is based on a number of north–south trunk routes, each operated by one of the five regional air carriers in Canada. They use jet aircraft and fly from a number of centres in the south to a few major northern airports on a regular basis in accordance with published flight schedules. From these airports smaller aircraft, often operated by smaller companies, fly to the surrounding settlements. Schedules are usually published for these flights also, but the companies are under no obligation to follow them. This pattern of supply is described as 'hub and spoke', the major airfields being the hub and the routes to the smaller settlements forming the spokes. It is a satisfactory system for the air carriers, allowing them to operate out of a few central locations, basing their aircraft and management there, and flying to individual settlements as conditions and traffic warrant. The 'hub and spoke' concept is not always in the best interests of the communities that are served, as the flights, for which the user has to pay, can be much longer than direct flights. For instance, settlements lying to the south of the hub are overflown on the northbound trunk flights, while two settlements can be near to one another geographically but widely separated in practice if they are served from different hubs. Traffic between two settlements served from the same hub often has to go through the hub, which can entail expensive delay there, and may be much longer than a direct flight.

The situation that existed at Pond Inlet until recently illustrates these problems. For many years Pond Inlet was supplied only from Resolute Bay, though there is a major airfield farther south through which the route would have been 500 miles (800 km) shorter. The Eskimos of Pond Inlet are very closely related to those living at Igloolik, less than 300 miles (500 km) away, with scarcely a man or woman who does not have a brother, a sister, a parent, or a child living there. Igloolik was, however, served from a different hub and a visit entailed flying through Resolute, Frobisher, and Hall Beach. This meant four flights in three or four different aircraft, with a good possibility of an extended delay at the traveller's expense at each change. The flight is 280 miles (450 km) for a crow, but it was over 1800 miles (2,800 km) by the air routes.

Another way in which the Air Transport Committee is able to assist the profitability of the carriers is by separating civil and military air services even where both are provided by civil carriers. As a result, a settlement in the north can have two quite distinct services, one for military and the other for civil use. A less expensive and more frequent service could be provided by combining the two. A further disadvantage is that the absence of integrated services encourages large industrial companies and government agencies to operate their own aircraft rather than use established and experienced carriers, aggravating the situation and making flying less safe. Current practices are difficult to justify in the context of energy conservation, and a strong case could be made for rationalization.

The main air routes to the north are from Montreal to Goose Bay, serving Labrador; from Montreal to Fort Chimo, Frobisher, Hall Beach, and Resolute, serving New Quebec, Baffin Island, and Foxe Basin; from Winnipeg to Churchill and Resolute, serving the Keewatin District; from Edmonton to Yellowknife, Cambridge Bay, and Resolute, serving the Western Arctic; from Edmonton to Inuvik, serving the Mackenzie Valley; and from Edmonton to Whitehorse, serving the Yukon. Most settlements in the north can now be reached from the south within two days, subject to weather. East–west routes in the north are much less developed, as there is little traffic.

Following the Second World War it was expected that airfields in the Canadian north would play an important role as refuelling points for international polar flights. The increased range and size of aircraft, however, soon made it more economical to fly direct, landing in northern Canada only if bad weather or emergency required it.

Trade

After the Second World War virtually all trade in the smaller settlements of the Northwest Territories and New Quebec was in the hands of the

Hudson's Bay Company. In the late 1950s, with the strong encouragement and support of the government, a number of co-operatives were organized by Eskimos and Indians. The concept of co-operation in enterprises was readily accepted by communities in which the people had been accustomed to help one another and to share the animals they hunted. At first producer co-operatives, engaged in such activities as carving in soapstone and fishing for char, were established. From there it was an easy step to set up consumer co-operatives importing and selling goods from the south. Many settlements now have stores operated by both the Hudson's Bay Company and a local co-operative, both selling very much the same range of goods and both purchasing fur. The Hudson's Bay Company has the advantage of long experience not only in operating trading posts but also in purchasing and shipping merchandise, and often attracts the lion's share of the trade. It is confident in its ability to compete with co-operatives alone, but not when they are supported by the government.

The main weakness in the co-operatives has been management, of which the native people have had little experience, and many have had serious financial problems. Sometimes they have attempted to rectify this by employing white managers. In 1972 the co-operatives in the Northwest Territories organized a central body, the Canadian Arctic Co-operative Federation, for bulk purchasing, arranging management training, and other matters that can be effectively handled in this way. In New Quebec La Fédération des Coopératives du Nouveau Québec, established in 1967, plays much the same role but has discouraged the employment of white managers.

The co-operatives have certainly helped to introduce the native people to the methods of the southern economy. In some places, too, they have moved into new activities such as providing municipal services, construction, and operating tourist lodges. In recent years individual Eskimos have begun to set up small private businesses dealing in a restricted range of merchandise or providing some service, such as the repair of over-snow vehicles, and catering to the needs of tourists, and this trend may continue, possibly at the expense of the co-operatives. Several settlements have also seen the appearance of white entrepreneurs who operate small enterprises and whose familiarity with commercial practices and lack of family obligations can give them a competitive advantage over native-run businesses.

Native land claims and environmental issues

In 1960 few people in Canada would have considered native land claims to be a factor of any significance in relation to economic development in the north. No doubts were ever raised of the need for development or that it would be to everyone's advantage, and not least to that of the native people. In fact the benefits it would bring to the native people were often used as a compelling, though unsupported, argument for promoting development that might not be economically sound. The north was considered an unoccupied land where anybody was free to seek and exploit resources with no need to inform, let alone consult with, those who had lived there throughout historic times. There was some justification for this attitude. By southern standards the land was practically uninhabited, and the few native people made no use of the resources sought by the modern developers. The Eskimos, in particular, had always appeared to welcome newcomers to the north and had often seen their interests as being complementary rather than competitive.

The situation has now changed to one of confrontation, with organized native groups opposing development, at least until the government has recognized their proprietory and cultural rights. Many reasons have contributed to this reversal of attitude. The most persuasive has probably been the scale of the invasion of the north by oil exploration companies. It became evident to the native people that this represented a threat to their whole way of life and that they could easily lose all control over their own future. Many of the younger natives had learned about southern ways and they were able to form aggressive political organizations and to represent the views of their people to southern audiences. They also had the example of Alaska, where the native people had managed to delay development until a land settlement had been reached, and where the land settlement was itself a recognition that the native people had rights that were of considerable value. At the same time public opinion in southern Canada was becoming more receptive to the views of the native people, partly because of the growing realization that their interests had been completely neglected in the past, and partly because of general disenchantment with the development ethic.

Another issue that has become increasingly important is the protection of the environment. Exploration companies had been allowed to do as they pleased in the north. Vehicles could travel without restriction leaving tracks across the tundra, camps could be abandoned, fuel drums left where they had been emptied, lines cut through the bush, and seismic charges exploded in lakes and rivers and at sea. Activities of this sort led to protests from two groups. The native people complained that their livelihood was being affected, particularly by disturbance of the animals they hunted. Biologists and other scientists drew attention to the fragility of the tundra, where degradation of the insulation provided by the surface layer led to thawing of the permafrost so that, for instance, vehicle tracks could increase in size year after year. They also pointed out that little was known of northern ecosystems, which appeared to be in delicate balance and peculiarly sensitive to change. The north was one of the few remaining wilderness areas; it was irresponsible to allow activities while so much had yet to be learned of their effects on the environment. These views were reinforced by the growing evidence of widespread arsenic and mercury poisoning in the north and the presence of DDT and other synthetic chemicals in northern animals. The government responded by drawing up land-use regulations designed to control and limit the damage to the environment caused by developmental interests, but they were criticized as inadequate, directed towards short-term palliatives instead of long-term solutions, and difficult to enforce, especially in their application to the activities of government agencies.

The proposed Mackenzie Valley Pipeline became the issue on which the questions of native rights and environmental protection focused. The scale of the project would be huge, its cost at least $9.5 billion, and it would without doubt affect the north profoundly. It was seen by the developers and many of the white immigrants from the south as the answer to northern social and economic problems and the means to make them all rich. From the first, however, the majority of the native people opposed the project, some because of concern about the direct effect it would have on their lives, others because they recognized that authorization by the government of a project of this scale would prejudice the case for native rights to the land.

This was the first time in Canada that the beneficial effects of northern development had ever been seriously challenged, and it soon became evident that many of the assumptions on which development had hitherto been based should be examined. The government therefore set up the Mackenzie Valley Pipeline Inquiry (the Berger Commission) to determine the social, environmental, and economic impacts of the pipelines that were being proposed and to report on the terms and conditions that should be imposed to moderate these impacts. Hearings were held over a period of two years in all the settlements that might be affected, as well as in the south. The native people seized this opportunity to present their case to the Inquiry, and in this way to the whole of Canada.

The report of the Inquiry recommended that no pipeline be built across the northern Yukon Territory owing to the damage to the environment that would result, and that a Mackenzie Valley pipeline should be postponed for ten years to allow time for the settlement of native land claims and for taking measures to reduce the social costs. It also expressed the opinion that a pipeline across the southern Yukon would not be open to the same objections as a pipeline across the northern Yukon.

The Inquiry has ensured that government decisions regarding pipelines in the Mackenzie Valley will be made in the light of much more complete and better-balanced information than had existed. Its findings will also be a powerful influence on the consideration of other proposals for northern development.

Administration and social conditions

At the end of the Second World War, the Northwest Territories was still governed by a commissioner, who was the deputy minister of the department responsible for administering the north, and a council of six appointed members, all federal civil servants living in Ottawa. The first change in this pattern came in 1947 with the appointment of a local resident to the Council. The evolution of the Council progressed gradually in stages until by 1964 it had four elected members from the Mackenzie District, and five appointed members, only one of them a civil servant.

It was not only in representation on the Council that the Mackenzie District differed from the rest of the Northwest Territories. Development was proceeding much more rapidly along the Mackenzie Valley, where mining had been established for many years and where a large proportion of the population were immigrants from southern Canada. They lived mainly in the towns, especially Yellowknife, and had little contact with the Indians, most of whom were hunters and trappers in the bush. The white immigrants were used to the ways of southern Canada which they wanted to introduce into the north, and they provided the only political voice in the territory.

The Keewatin and Franklin Districts were also changing, but in different ways. It had needed the Second World War to bring the Arctic to the attention of the Canadian public, which was surprised and disturbed to find that over a large part of the country educational and health services were to all practical purposes almost absent, and that the native people were living in primitive conditions that appeared much less tolerable to southern Canadians than they did to the native people themselves. Great efforts were made in the succeeding years to improve standards, with cost a secondary consideration. The course followed was administratively simple; the government decided what was needed according to southern models, and provided it in good measure. The alternative of consulting the native people themselves, shaping social development in accordance with their wishes, and involving them directly in the provision as well as the receipt of welfare, was suggested, but never seriously considered. The thinking that pervaded the administration was well illustrated in the field of education in the Arctic. The highest priority was placed on the provision of physical plant, and schools were built throughout the north. The schools were staffed by teachers imported from the south and teaching provincial curricula. Education was seen essentially as the instrument of assimilation and the way to prepare the native people for participation

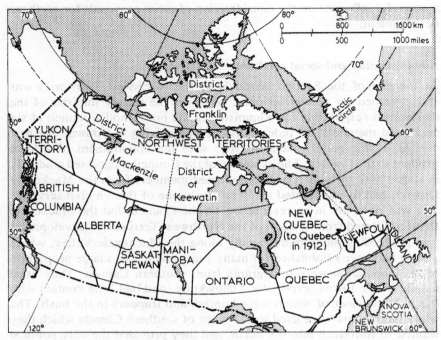

Map 13. Canada. Administrative divisions

in the rapid economic development that was confidently expected; native languages were never used in the schools; and no effort was made at the time to train Eskimos as teachers. It was a very paternal system, and father was rich rather than understanding.

In health services a vigorous attack was made on tuberculosis, which was widespread. Treatment involved long periods in southern hospitals, often a heart-breaking experience for the closely knit Eskimo families, but justified by its medical success. The establishment of schools and nursing stations in the main settlements induced many of the Eskimos to move there from the small hunting camps, making it more difficult for them to hunt and trap. There was no other significant source of income at these settlements, apart from wage employment in service occupations in white establishments, so relief became widespread, and the Eskimos depended increasingly on the government.

With development following such divergent paths in the Northwest Territories, the government proposed in 1963 to divide the area into two separate Territories, Mackenzie in the more advanced west, and Nunassiaq in the predominantly Eskimo Eastern Arctic. It was thought that the governments of these Territories would be able to respond more effectively to the very different needs in their areas than could a single territorial government attempting to cover an area equal to a third of all Canada. The government's proposal was, however, strongly opposed, partly because it was widely interpreted as an attempt to divide and continue to rule, and partly because of the proposed boundary between the two new Territories, which was difficult to justify as it cut across the area inhabited by Eskimos in order to include certain potentially valuable resources in the proposed Mackenzie Territory. The proposal was therefore dropped and an Advisory Commission on Development of Government in the Northwest Territories (the Carrothers Commission) was appointed. The Commission travelled widely throughout the north in an attempt to determine local views but it was not possible to explain the issues and receive considered opinions in what was of necessity a series of hurried visits. The major recommendations of the Commission, made in 1966, were that the Northwest Territories should remain undivided for ten years and the question should then be reopened, and that the seat of the territorial government be moved without delay from Ottawa to Yellowknife. In the meantime the Council had been increased by adding elected members from the parts that had been unrepresented.

So far as the Eastern Arctic was concerned, the move to Yellowknife did not bring government any closer, but lavish funding by the federal government of the territorial government allowed a continued expansion of services at the main settlements and greatly improved housing there. This led to further concentration of the Eskimo population, making them more dependent on the government. With little control over their own future and often unable to understand the changing administrative policies and procedures which determined their day-to-day life, many Eskimos became demoralized, especially at the major centres where white influence was greatest and alcohol easily obtainable.

In the Mackenzie Delta the increased oil exploration activities following the discovery of the Prudhoe Bay field in Alaska began to interfere directly with hunting and trapping, alerting the native people there to the magnitude of the threat to their whole way of life. They were aware of the progress of negotiations on land claims in Alaska as many of them had near relatives in Alaska, and they formed the Committee for the Original Peoples Entitlement to determine and promote their own claims. This development was followed by the establishment in August 1971 of Inuit Tapirisat, as association of all Eskimos in Canada, with the objectives of strengthening the Eskimo culture and language, informing the Eskimos of their rights, formulating their claims, and representing them in negotiations with the government. They were assisted by the Indian-Eskimo Association of Canada, which consisted mainly of white Canadians concerned at the poverty among the native people. The Indians of the Yukon and of the Northwest Territories had formed brotherhoods for much the same purposes, and all these organizations applied for and received financial support from the federal government.

This growth of native organizations in recent years has overshadowed any other political developments in the Territories. The Council of the Northwest Territories has, like that of the Yukon Territory, become a fully elected body, and its membership has been increased to fifteen. In both Territories, elected members head certain administrative departments. In the Northwest Territories Council the native people have gained a majority of the seats. The territorial councils have, however, failed to gain the confidence of the native peoples, who tend to see them as the instruments of developmental interests rather than a means of achieving reforms that will benefit them. The native organizations, on the other hand, have become vocal pressure groups, and their growth has been followed by a number of administrative changes to meet the wishes of the native people. Among these are the introduction of native languages in the schools and the provision of higher grades in local schools to avoid having to send young children away from home to boarding schools.

The acid test of the effectiveness of the native organizations will be the land claims settlements. The federal government is anxious to reach an understanding with the native people but is unwilling to consider their suggestion of a moratorium on development for a period of ten years. Many, perhaps a majority, of the native people will not be satisfied with a cash settlement, however generous, and want to ensure a continuing measure of control in the government of the land which they consider theirs. The events in the James Bay area have provided a blunt warning. Here the gigantic hydroelectric development that will transform the region was planned without any consultation with the native people, whose traditional way of life it will inevitably destroy. The James Bay settlement was agreed to reluctantly by the natives of the affected area, who had little opportunity to prepare their case, and was based largely on financial considerations. Inuit Tapirisat, along with other native groups, does not consider that the settlement represents adequate compensation for the rights it extinguishes, and has rejected it as a precedent.

In 1976 Inuit Tapirisat proposed the establishment of a new Territory within Canada, called Nunavut, roughly covering the districts of Keewatin and Franklin, where the native people would have the major role in the government and receive royalties on minerals taken from it. This proposal, which was later withdrawn for revision, bore some similarity to the establishment of Nunassiaq suggested in 1963 by the federal government, and the first official reactions were not unreceptive. The payment of royalties is less important than it might appear, as the north is heavily subsidized by the federal government. Canada has passed the stage when one area or sector of the population can be allowed to be either much richer or much poorer than another, and the form in which equalization payments are made does not really matter much. A serious objection to the Nunavut proposal was that it provided for the substantial funding in perpetuity of Inuit Tapirisat, a measure that holds the seeds of one-party government for the proposed Territory.

The Indian Brotherhood of the Northwest Territories has made a somewhat similar proposal for the establishment of a Dene Territory in the Mackenzie District. The problem is more difficult here because of the large white population, with very different interests, already resident in the Mackenzie Valley. The protection of Indian identity and culture in the Yukon, where they form only a small minority, will be even harder to achieve.

The future

Over the past century the forestry and mining industries have moved progressively farther north, at some times more quickly than at others, for the north as a marginal area is particularly sensitive to changes in economic conditions. But the long-term trend has been clear. Growing markets in the south, increasing prices, and depletion of more convenient resources have pushed back the frontier of development. The forestry industry cannot expand beyond the treeline, which in some areas it is now approaching. In the case of the mining industry the same factors that have applied in the past will in all probability continue, and mining will move farther into the north.

A number of new factors have, however, appeared and these will modify both the rate and nature of development. At the end of the 1960s petroleum exploration spread rapidly over the north and became the dominant commercial activity. Owing to disappointing results, drilling has declined sharply, but concern at the low levels of Canadian petroleum reserves will ensure that exploration continues. Most drilling has been carried out in areas which have been uninhabited and very rarely visited during the past century, using crews flown in directly from southern Canada. Supplies have also been sent in from the south by air or ship. As a result the direct effect of petroleum exploration on the native people of the north has been limited. It is rather the transportation of any production that could have wide repercussions. Oil discoveries have been meagre and there is no prospect at present of reserves in quantities that could not be exploited far more economically by sea transport than by pipeline. Natural gas presents a different picture. Reserves are still well below the threshold required for the construction of a pipeline, but after a few more seasons of drilling they might reach this level in the Sverdrup Basin. If, however, the gas is liquefied and transported by ship, and if the Prudhoe Bay gas is piped through Alaska, the effect of oil development on both the northern terrain and the northern people would be much less than previously expected, and the situation might not be far removed from the ten-year moratorium suggested by both conservationist and native groups and endorsed for a Mackenzie Valley pipeline in the report of the Mackenzie Valley Pipeline Inquiry. A decrease in developmental activity would tend to diminish the influence of these groups, whose political leverage depends to some extent on their capacity to delay and hence on how urgently industry requires the necessary authorizations.

Petroleum is not the only source of energy in the north. There are extensive coal deposits, but it will be very many years before they are likely to be worked, except possibly for local consumption. Uranium is of much more immediate importance and there is renewed interest in uranium prospecting in the northern Territories, a favourable area which has already had two uranium producers.

Another factor that could affect the north is the changing attitude towards development, which is no longer seen as its own justification. Conservationists, disturbed by interference with natural ecological systems and concerned by irreversible environmental changes and persistent pollutants, are making common cause with the native groups in demanding more stringent controls. The fact that many companies involved in resource development are owned or controlled by foreign interests gives these arguments a wider appeal.

Few of the northern natives are philosophically opposed to development. They naturally resent development which ignores or rides roughshod over their interests, but it would be a different matter if they were assured of a reasonable measure of control. It is up to the developers to adjust their plans and accommodate the wishes of the native people, and to convince them that they will benefit. Now that developers are beginning to recognize the need, they may succeed where they have hitherto failed.

There are other factors which will affect the nature rather than the rate of development. The scale of northern mines is increasing; they are capital-intensive, handle very large tonnages, use advanced technology, and are in no way dependent on the north for support. Only during construction are they major employers of labour. The influence of a mine could therefore be localized to its immediate vicinity, a policy that would have considerable appeal to many of the native people. At the same time improved air transport, the increased costs of living in the north, and the absence there of many of the amenities to which southern Canada has become accustomed, suggest an alternative to establishing residential towns in the north. It is becoming more attractive for men to be flown in to the north for short periods of intensive work, followed by a rest period in the south where their families continue to live. This practice, already followed in oil exploration, will probably be adopted more widely. The largest employers by far in the Territories are the federal and territorial governments; the great majority of public servants are white at present, but the proportion of Eskimos and Indians is certain to increase. Again the effect will be to limit southern influences, and reduce the size of the residential white population, a matter of considerable political significance. Under these circumstances industrial development of the north and the continuation of some form of traditional life based on renewable resources are not necessarily incompatible, and proposals along the lines of Nunavut become practical possibilities.

CHAPTER FOUR

Alaska: the evolution of a northern polity

Introduction

The European discovery of Alaska was made in 1732 by the Russians M. S. Gvozdev and I. Fedorov when they sailed across Bering Strait. (Here we leave aside the possibility, which is quite likely but unproven, that earlier Russian seafarers may have seen it about the middle of the seventeenth century.) Their achievement was long unknown, and hence the generally accredited discovery was that made in 1741 by Vitus Bering and the members of the second expedition sent from Russia to discover where Asia ended and America began, or if they were joined. It was of greater contemporary importance, however, that the survivors of Bering's expedition returned with some sea otter skins and information on the animal life of the newly discovered mainland, islands, and offshore waters. The eastward course of Russia's empire and the fur trade converted the passage from Kamchatka to Alaska into a busy sea lane, and in 1799 the Russian-American Company was granted a monopoly over all commercial enterprise and the governing of Russian America. Commercial penetration was preferred to a direct claim of sovereignty, partly because Britain was seen to have succeeded so well with the East India Company and the Hudson's Bay Company. The Company's peak activities extended as far south as Fort Ross in California, but by the mid-nineteenth century it was of little economic value to the motherland or its shareholders. The primary fur resources were severely depleted, sovereignty and monopoly claims were ignored by aggressive traders and whalers from New England, and the invasion of Russian America by the Hudson's Bay Company had been formalized by a face-saving lease over southeast Alaska. The colonial experiment had proved a failure, for at no time did Russian settlers reach 1,000, and there were no peasants to provide a local food base.

On 18 October 1867 Alaska was officially transferred from Russia to the United States, although it was not until the summer of 1868 that the US Congress could be persuaded to appropriate the $7.2 million purchase price. The acquisition was not a popular one, even supporters of the expenditure being convinced that the transaction had no commercial value but was justified either as a future strategic asset or a present act of friendship to Russia. The territory did prove to have commercial value, dramatically highlighted by the gold stampedes of the 1880s and turn of the century, and the current oil boom; and the advent of the air age and the Second World War underlined its strategic value.

The acquisition of Alaska also set in motion another process which has been an American tradition since the first expansions from the eastern seaboard and a dominant theme in Alaska's history after 1867, that of settlement and political development of new territories. Speaking on behalf of Alaska's development as late as 1915 a US Congressman from Minnesota gave this a classic statement:

> When the United States acquires domain over extensive tracts of territory, the duty devolves upon it not so much to exploit the natural resources for the benefit of the people of the States as to build there a civilization, to induce immigration and settlement . . . that homes may spring up and that that territory may contribute to the general strength and happiness of the whole Union.

Generous use of this type of political rhetoric succeeded in changing Alaska from 'Seward's Folly' in popular sloganmongering to 'The Last Frontier', and an effort was launched to replay the nineteenth-century 'Winning of the West' in a new geographical setting. In spite of application of homesteading and land programs, agricultural and settlement subsidies (culminating in the 1935 Matanuska Valley agricultural resettlement project), and the building of a railroad (1923) to 'open the land', the results were disappointing. Alaska lost population between the 1900 and 1920 census, and remained stagnant until 1940. Although population expanded dramatically and continuously after 1940, the number of farms fell from 623 in 1939 to 310 in 1968. In a 1937 report on the value of Alaska to the nation, the National Resources Committee concluded that the expected functioning of this process not only was unrealistic in this northern environment, but was a hopeless anachronism in an age of increasing urbanization and specialization.

Since the mid-twentieth century mark, however, Alaska has expanded rapidly in population, and in economic and political development. A contemporary American society and polity appears to be emerging in Alaska, as hoped for by the romantics and visionaries of the last century, but its character and the path of its evolution contradicts the blueprints of public policy and private dreams. For reasons of physical geography and market remoteness, Alaska could not follow the evolution of an agricultural-based society, but it has found a path leading directly towards development of a twentieth-century urban society. The current popular belief that the future rather than the past will set the Alaskan patterns of development is implied in the latest Alaskan slogan, 'North to the Future '. But the process by which this will be achieved is an evolutionary one and there is a present danger that popular and official views may be based simply upon the exchange of one set of myths concerning a pioneering past for another set inspired by futurology.

Geographical background

The Alaska Boundary Settlement of 1903 between the United States and Great Britain established the political boundaries. So defined, Alaska is a large peninsula bounded by the Arctic Ocean on the north, the Bering Sea on the West, the Gulf of Alaska and the North Pacific Ocean on the south, and the 141st meridian and the crest of the Coast Range on the east (Map 14). The total land area is 586,400 sq miles (1,520,000 sq km), stretching out between latitudes of 51° and 72° N and the meridians of 130° W and 173° E, and containing four time zones. The 1964 Convention on the Continental Shelf (submarine areas adjacent to the coast to a depth of 200 m) added 550,000 sq miles (1,430,000 sq km) of sea bed and its resources to the land and resource base of Alaska, 65 per cent of the total United States' continental shelf.

Alaska is not a single homogeneous region but several distinctive regions. The Pacific Mountains System defines the northern and eastern coasts of the Gulf of Alaska and the North Pacific Ocean and the southern border of Alaska. It is a continuation of the continental system swinging northward through coastal British Columbia into Alaska as the Alexander Archipelago and Coast Range of southeastern Alaska, and sending off two spurs as it continues west and southward in an arc across the Gulf of Alaska. The coastal spur contains the St Elias Range with peaks reaching up to 19,000 feet (5,800 m) above sea level, the Chugach and Kenai Mountains, and Kodiak Island. The main spur forms the Alaska Range culminating in 20,300-foot (6,200 m) Mt McKinley, and continuing south and westward as the backbone of the Alaska Peninsula and the Aleutian Islands. Between these two spurs lie the Inside Passage and other water-ways of southeast Alaska, the Wrangell Mountains, Copper River

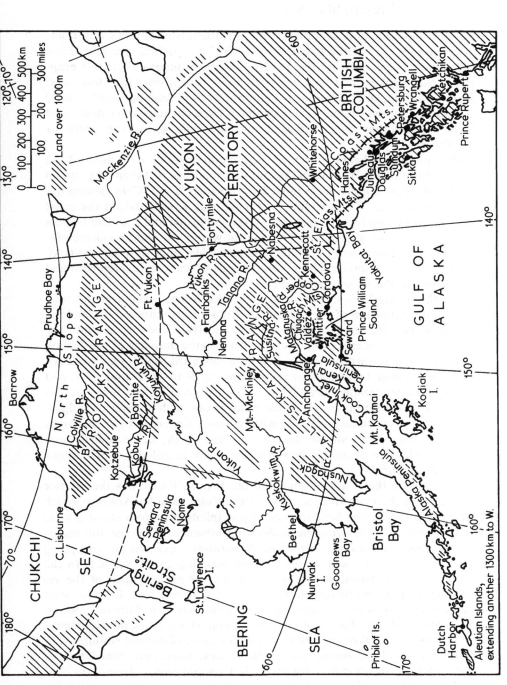

Map 14. Alaska

plateau, the Talkeetna Mountains, the Susitna lowlands and Cook Inlet. The region is part of the Pacific 'rim of fire' and includes several active volcanoes. It has been subject to frequent and at times destructive eruptions, earthquakes, and tsunamis. The most recent were the 1964 Good Friday earthquake centered near Valdez and felt throughout south central Alaska (8.3 on the Richter scale) and the related tsunami which destroyed Valdez and other coast villages, and did damage as far south as California.

Precipitation is heavy: about 60 in (150 cm) annually over most of the region, but ranging from 14 in (35 cm) at Anchorage to 150 in (375 cm) on the coast. Mean temperatures range from about 55 °F (12 °C) in the summer to about 32 °F (0 °C) in winter. As far west as Kodiak Island the coast is covered by rain forests, and the Alaska Peninsula and Aleutian Islands contain extensive grasslands. The coastal waters are rich in marine life, principally salmon, and the land contains a wide range of minerals: gold, coal, and copper having major roles in past economic development and petroleum in the present and future.

Northern Alaska consists of the Brooks Range, which is the final northward extension of the continental Rocky Mountain system, and the North Slope, drained by northward flowing rivers and streams into the Arctic Ocean. It merges to the southwest into the drainage of the Kobuk River and the Seward Peninsula. Precipitation is low (6 in (15 cm) or less), average summer temperatures range from 32 °F to 55 °F (0–12 °C), and average winter temperatures from 3 to 16 degrees below zero (−20 to −25 °C). The entire region is in the tundra zone with a cover of sedges, mosses, lichens, small brush and willows, most underlain by permafrost. Caribou, polar bear, smaller fur-bearers, game fish, whale, walrus, and seal are of some economic importance. Today petroleum overshadows all other resources.

In between the Alaska Range and the Brooks Range lies the interior plateau and the main river systems of the Yukon and Kuskokwim which drain into the Bering Sea. In general precipitation is from about 4 to 20 in (10–50 cm). Cut off from the moderating influences of the sea, this region has the greatest seasonal contrast in temperatures, extreme lows dropping to −70 °F (−55 °C) and extreme highs reaching 100 °F (40 °C). In the interior, boreal forest occupies the main river basins, most of the remainder of the land being tundra underlain by permafrost. Big and small game mammals, sport fish, and migratory wildfowl occur. Minerals, principally gold and coal, have been important in past economic development.

The total natural environment and the location of natural resources, in interaction with economic and social forces, have determined human habitation and development patterns. Man's occupation and use of the land and the addition of social features (roads, towns, etc.) have called for

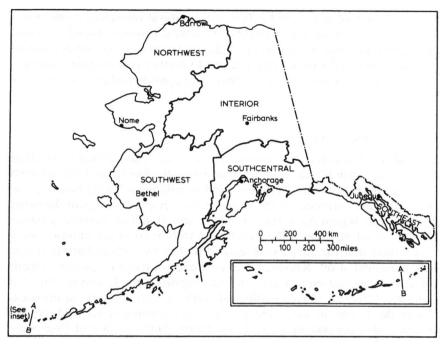

Map 15. Alaska. Socio-economic regions by 1970 census divisions

the division of Alaska into socio-economic regions. This discussion will use the five regions indicated in Map 15. These units do not have any governmental or administrative standing, but have been widely used for planning, research, and statistical purposes.

Resource base

Alaska's development before the Second World War can be traced in patterns of exploitation of its natural resource base. European explorations of the Pacific northwest during the eighteenth century prepared the way for the extension of the Russian and the British fur trades into what is now Alaska, and launched its colonial development. A pattern of destructive exploitation was set by whaling (1847–53) and the harvesting of fur seal, sea otter pelts (1768–1911), and a variety of land furs. After the transfer of Alaska to the United States (1867), the beginnings of the canned salmon industry (1878–84) and the major discoveries of gold (1880–1906) provided the base for an expanded colonial economy. Between 1911 and 1938 copper ore from the Kennecott mines made a further major contribution (from 1915 to 1928 the value of copper production exceeded that of gold), and a few other natural resources made minor contributions. There was a decline in natural resource-based activities during Alaska's period of defence build-up (1940–58), but the decades of the 1960s and 1970s have seen a return to further resource development in forest products and petroleum, and an expansion and diversification of fisheries.

Non-renewable resources

In 1968 the discovery was announced of an oil- and gasfield at Prudhoe Bay on the North Slope, containing an estimated five to ten billion barrels (0.7–1.4 billion tonnes) of crude oil and 26 trillion cu ft (740 billion cu m) of natural gas. Estimates of the ultimate recoverable reserves of the entire North Slope region have since ranged from 50 billion barrels (7 billion tonnes) of oil and 300 trillion cu ft (8.5 trillion cu m) of natural gas to speculation that the range may extend to 100–150 billion barrels (14–20 billion tonnes) of oil. Knowledge of existence of these resources extends back to investigations of oil seepages reported by Eskimos. The US Geological Survey started geological work in 1901, private oil firms explored the region in 1921, and the federal government set aside 37,000 square miles (96,000 sq km) of the North Slope as Naval Petroleum Reserve No. 4 in 1923.

Discoveries at Swanson River on the Kenai Peninsula in 1957 were also preceded by exploratory drilling in 1902, and modest commercial production of crude oil was realized in the Katalla district near Cordova in the 1930s. The reserves of the Cook Inlet oil and gas region are estimated at 1.5 billion barrels (200 million tonnes) and 5.13 trillion cubic feet (146 billion cu m) of natural gas.

Although reserve estimates are not available, over 60 million acres (240,000 sq km) of possible petroleum provinces exist south of the Brooks Range in the lower Yukon and Kuskokwim region, Bristol Bay and Alaska Peninsula, and lower Cook Inlet. Stimulated by the national drive for energy self-sufficiency (Project Independence), the US Secretary of the Interior listed on 25 May 1974 nine continental shelf areas for leasing for exploration and development: Gulf of Alaska (eastern Gulf, Kodiak Island and Aleutian Island shelves); Bering Sea (St George Basin, outer Bristol Basin and Norton Basin); Chukchi Sea (Hope Basin); Beaufort Sea; and lower Cook Inlet (south of 60° N).

The most extensive known deposits of coal are north of the Brooks range extending eastward from Cape Lisburne for about 300 miles (500 km). Inferred reserves are about 19 billion short tons (17 billion tonnes) of bituminous coal and 100 billion short tons (90 billion tonnes) of sub-bituminous and lignite. Smaller fields are at Nenana north of the Alaska

Range, Matanuska near Anchorage, Beluga, Susitna, and Bering River near Cordova. Development has been limited to locations along the Alaska Railroad right of way and was at its peak when the railroad and military installations burned coal for heat and energy. Japanese investigations for coking coal have centred in the northern and Bering River fields.

Alaska's known reserves of gold exceed 100 million troy oz (3,110 tonnes), which is about four times the total production since 1880 from large lode mines near Juneau, from placer operations near Fairbanks, at or near Nome, and on the Kuskokwim River, and from other smaller operations scattered throughout Alaska. Next to gold, copper has been the most important metal produced in Alaska, past operations being located at the large Kennecott mines in the McCarthy district and smaller mines in Prince William Sound and southeastern Alaska. Copper production extended over almost three decades (1911–38). The largest known copper deposit is at Bornite in the Kobuk River area, with more than 500 million short tons of ore containing 1.2 per cent copper, followed by 200 million short tons of 0.4 per cent copper at Orange Hill in the Nabesna area, and 50 million short tons of 1 per cent copper at Sumdum in southeast Alaska. Other small deposits in Prince William Sound, Cook Inlet, and southeast Alaska total about 50 million short tons of 1 to 2 per cent copper.

Alaska has several billion tons of 10–35 per cent iron ore, with huge deposits at tidewater in southeast Alaska (Klukwan near Haines, Port Snettisham, and Union Bay), the west side of Cook Inlet and the Nushagak Basin at Bristol Bay. All are of low grade with a titanium-to-iron ratio between 1:5 and 1:10.

Tin, usually associated with tungsten and frequently with beryllium, is found in deposits that extend eastwards from the western portion of the Seward Peninsula into interior Alaska. The most important tin deposit is at Lost River in the western portion of the Seward Peninsula. Tin, associated with placer-gold production, has also been produced from the Hot Springs mining district, and has been noted in most of the interior Alaskan placers. Lesser tin deposits have been worked sporadically for many years, but no major tin placers comparable to those of southeastern Asia have been discovered.

Occurrences of lead, zinc, and silver deposits are widely scattered, and most production of these minerals has been as a by-product of gold-mining. Southwest Alaska is one of the world's important mercury provinces, with about thirty mines and prospects. Reserve estimates are not available, but developed deposits have run about 35 to 45 lb of mercury per short ton (18–22 kg per tonne), several times the United States average. Since 1938 the Goodnews Bay platinum placers near Bristol Bay produced about 27,000 oz (770 kg) annually until the 1960s, after which annual production fell to about half that amount. This occurrence is so far unique in Alaska.

Stibnite, the principal ore of antimony, is widespread in the eastern and central interior region and the Seward Peninsula. Deposits of the beryllium ore, chrysoberyl, at Lost River total 2 million short tons (1.8 million tonnes) containing 0.15 to 0.54 per cent beryllium and averaging 50 per cent fluorite. There are estimated to be 420,000 short tons of reserves of chromite near Kachemak Bay, which has produced about 30,000 short tons, and potential for locating further reserves is favourable. About 170,000 lb of tungsten (77.5 tonnes) were produced during the Second World War from the Fairbanks area and southeastern Alaska, and new production is possible in connection with tin developments planned at Lost River. There is a reserve of 23 million short tons of nickel ore in southeastern Alaska and a uranium mine operated there in 1957–65 and 1971–3.

Of the non-metallic minerals, sulphur occurs in association with recent volcanic action on the Aleutian Islands and in iron sulphide deposits in Prince William Sound and southeast Alaska. Limestone of high purity is found on the islands of southeastern Alaska, and in erratic and lower grade deposits along the Alaska Railroad and Kenai Peninsula. Asbestos has been found at Kobuk, Forty Mile, and in southeastern Alaska, where barite deposits have been developed in connection with petroleum development needs on the North Slope. Graphite is found on the Seward Peninsula (65,000 short tons of high grade and 300,000 short tons of low grade) and nephrite and jadeite in the Kobuk region. Large quantities of sand and gravel have been used in construction.

Renewable resources

Alaska contains an estimated 123 million acres (500,000 sq km) of forest lands (16 per cent of all forest lands in the United States), of which 28.8 million acres (110,000 sq km) are considered commercial forest. The rain forests of the coastal regions, extending from southeast Alaska across the Gulf of Alaska to Kodiak Island, cover some 16.5 million acres (63,000 sq km), of which 5.75 million acres (23,000 sq km) are commercial. The annual allowable cut under sustained yield practices is estimated at 1.1 billion board feet (2.6 million cu m) with the possibility of considerable expansion on a second-cut basis. Present harvest is about half this amount. Of the commercial stands, western hemlock accounts for 64 per cent of the volume, Sitka spruce 29 per cent, and cedars 7 per cent. Two-thirds are of pulp quality and the remainder suitable for timber and plywood veneer. The interior forests are located along the major river systems and cover an estimated 106 million acres (425,000 sq km), of which 22.5 million acres (90,000 sq km) are commercial. Forest inventories are not complete and reliable estimates of probable allowable cut are not available.

The dominant species are white and black spruce, with birch, aspen, and cottonwood as secondary species. Most of this is of pulp quality, although some of the birch appears suitable for veneer production.

Fisheries, primarily salmon, provided the staff of life for most of Alaska's indigenous people and until the recent expansion of petroleum represented the largest source of commercial value from natural resources. Despite sharp decline in salmon catches since the peaks of the 1930s and 1940s, Alaska continues to lead all other states in value of its commercial catch and fisheries products. For the five-year period 1970–74, the annual commercial catch landed in Alaska averaged 236,000 tonnes (44 per cent was salmon; 46 per cent shellfish; and 10 per cent other fish, principally halibut and herring). These fisheries are located in the coastal waters of Alaska, the northeastern Pacific, and the Bering Sea, and are more fully treated in Chapter 7. Landings are made by Alaskan fishermen at ports along the entire southern coast from Ketchikan to the Aleutian Islands, Bristol Bay, and to a considerably lesser extent the remainder of the Bering Sea coast. Outside territorial waters salmon, halibut, crab, and other species are caught by foreign fishermen in competition with Alaskans, with significant effect upon the stock of salmon and other fish which spawn in Alaska. Calculations based on estimates of the fish stocks of these waters indicate that Alaskan salmon, halibut, and most species of shellfish are already exploited to the allowable limits for sustained yield and sometimes beyond. Although Japanese, Russian, and other high-seas fisheries fleets have harvested other flatfish and demersal fish, these stocks appear to be sufficient to permit further expansion of Alaska-based fisheries.

Fur, the first of Alaska's natural resources to attract European and North American exploitation, still yields a significant harvest. Trapping of land mammals such as mink and beaver accounts for about one-quarter of contemporary value of the total fur industry in Alaska, hair seals about 10 per cent, and the United States' fur seal share between 60 and 70 per cent. The fur-seal resource, consisting of about 1.75 million seal which seasonally come to the Pribilof Islands to breed, is managed and harvested by the United States.

The annual run-off of Alaska's streams is estimated to be 300 million acre-ft or 10 in (25 cm) depth from the entire State. There are extensive areas of ground water. Water supplies are more than adequate to meet all foreseeable municipal, industrial, or irrigation needs without impairing the production of hydroelectric power. This has given rise to proposals to pipe Alaskan water through Canada and down to southern California and southwestern areas of chronic water shortage. Alaska abounds in hydroelectric power potentials of which only a few minor projects have developed. Interest has focused on the power giants – the 5,040 MW Rampart project on the upper Yukon River, and others of up to 3,500

MW at Wood Canyon on the Copper River, Woodchopper on the upper Yukon, the Yukon–Taiya project in southeast Alaska and Yukon Territory, and the upper Susitna River project. The Yukon–Taiya project appears to be the most feasible from the Alaskan point of view, all the others presenting serious conflicts with other resource values.

Alaska's great size gave rise to expectations that it could develop a correspondingly significant agricultural economy. The most optimistic appraisals, however, indicate that only 1 per cent of Alaska's land area has soils suited to cultivation and an additional 2 per cent might be usable for grazing sheep and cattle. Most of the presently accessible potential farmland is located in the central Tanana Valley of interior Alaska and the Matanuska and Susitna Valleys in south-central Alaska. Climate and economic factors, however, have limited actual development. According to a 1967 survey, 27,266 acres (108 sq km) of land were tilled, 18,736 acres (75 sq km) were idle crop land, and 209,178 acres (830 sq km) were used as pasture or range lands. Reindeer grazing areas in Alaska are in the Seward Peninsula and Nunivak Island. In the mid-1930s an estimated 600,000 reindeer were herded or grazed in Alaska, but by 1969 only 31,000 head were to be found on range areas totalling 22,000 square miles (57,000 sq km). Potential exists for commercial expansion, but efforts have been limited to a few native-owned undertakings, the most successful being on Nunivak Island. Domestication of muskoxen is still in an experimental stage.

Alaska's game mammals, sport fish, and scenery provide the basis for a growing tourism and provide attractions to Alaska's urban residents.

Population

Patterns of population change and distribution have been determined by four major development patterns over the past two hundred years. The first was the aboriginal pattern of occupation and resource uses which, although eclipsed by subsequent development, has survived in certain essentials to reappear today as a vital part of the contemporary native Alaska political movement and of the land ownership changes implicit in the native claims settlement legislation.

At the time of the first European contacts the southeast region had an estimated 10,000 Tlingit and 1,800 Haida Indians who were part of the aboriginal culture of the northwest coast of North America. Here the mild climate and abundance of readily harvestable marine resources provided the wealth and leisure for elaboration of a relatively rich culture and sophisticated social system. Western Eskimos and Aleuts inhabited the shores of the Arctic Ocean, Bering Sea, and Gulf of Alaska, following a more spartan existence based on marine resources, and penetrated the

Arctic inland as hunters of caribou. Estimates of the western Eskimos put about 7,200 along the Arctic coast, 600 on St Lawrence Island and the Bering Strait, 11,000 along the Bering Sea coast and delta lands, and 8,700 on Kodiak Island and along the Gulf Coast. An estimated 16,000 Aleuts followed a maritime existence along the chain of the Aleutian Islands and on the Alaska Peninsula. Some 6,900 northern Athapascan Indians were scattered in small groups throughout interior Alaska, and in the Cook Inlet and Copper River regions.

The Russian and initial United States occupations of Alaska were dominated by a traditional colonial economy of highly specialized natural resources exploitation for the benefit of distant markets. The expansion of commercial fisheries and gold-mining in the last decade of the nineteenth century is reflected in the rise in non-native population from the few hundred during the Russian and initial United States periods of the fur trade and whaling to 39,025 in the 1909 census (Table 9). The period of stabilization and stagnation of the basic economy of gold, fish, copper, and furs is reflected in the decline between 1909 and 1929 in non-native population, and the effects of the Great Depression outside Alaska, which stimulated increased gold production and a 'return to the land', in the regaining of the 1909 non-native population level by 1939.

The native population dynamics were the result of natural forces of fertility and mortality, until recently there being very little migration out of Alaska and relatively little within Alaska. These biological forces were modified by the disruptive contact between a self-sufficient subsistence culture of an aboriginal people and specialized and exploitive colonial forces and the depredations of unfamiliar diseases. The accelerating increase in native population, starting in the late 1920s and assuming explosive proportions in the 1950s, reflects generally successful programs of public health and welfare in keeping people alive, and an absence of birth control.

Colonial Alaska was eclipsed between 1940 and 1942 by the coming of the Second World War to Alaska. For the next two decades, income generated by all of Alaska's resource extraction and processing industries combined was exceeded by military payrolls alone. By the 1960s the defence economy had levelled off, and its employment and income-producing capacity declined. Subsequently the basic economy began to reflect the State's increasing importance in intercontinental air traffic, and as a source of a broader range of natural resources for domestic and foreign markets.

Expansion in population following the 1939 census largely reflects fluctuations in defence personnel from 524 in 1939 to 154,000 in 1943 and 50,000 in the 1950s. From the 1950s these personnel have been accompanied by an equal number of dependants, several thousand civilian

Table 9. General population trends in Alaska, 1740–1970

Year or date	Total No. of persons	Trend[a]	Native No. of persons	Trend[a]	Non-native No. of persons	Trend[a]
Circa 1740–80	62,200		62,200	100.0	...	...
1839	39,813	13.2	39,107	62.9	706	0.3
1880	33,426	11.1	32,996	53.0	430	0.2
1890	32,052	10.6	25,354	40.8	6,698	2.7
June 1900	63,592	21.0	29,536	47.5	34,056	13.5
31 Dec. 1909	64,356	21.3	25,331	40.7	39,025	15.5
1 Jan. 1920	55,036	18.2	26,558	42.7	28,478	11.3
1 Oct. 1929	59,278	19.6	29,983	48.2	29,295	11.6
1 Oct. 1939	72,524	24.0	32,458	52.2	40,066	15.7
1 Apr. 1950	128,643	42.6	33,863	54.4	94,780	37.7
1 Apr. 1960	226,167	74.8	43,081	69.3	183,086	72.8
1 Apr. 1970	302,647	100.0	51,128	82.2	251,519	100.0

[a] Number of persons expressed as percentage of maximum for each series.
(Sources: Data for 1740–80 based on estimates in J. W. Swanton, *The Indian Tribes of North America*, (1952) and W. H. Oswalt, *Alaskan Eskimos*, (1967). For 1839 based on estimates by Veniaminov and others in 'Resources of Alaska', *10th Census of the United States*, 1880, vol. VIII, pp. 36–8. Other data from US Bureau of the Census reports 1880 to 1970. The 1 April 1970 total population from PC(1)A3 issued May 1971. Native and non-native for 1970 as tabulated from census tapes in 'Age and sex characteristics of Alaska's population', *Alaska Review of Business and Economic Conditions* (ISEGR), March 1972. The 1 April 1970 data corrected for undercounts.)

employees of the Department of Defense (and their dependants), and a fluctuating labour force of supporting workers related to the construction and maintenance of the defence establishment. The decline in military personnel to 25,348 in 1975 would be expected to be reflected in similar behaviour of the civilian economy, but the total population has continued to rise since 1960, reaching an estimated 404,624 by 1 July 1975. New developments in petroleum (including construction of the trans-Alaska pipeline) and forest products, and the expansion of civilian government services following statehood, created more jobs and a new in-migration. Family-planning programs and the legalization of abortion in 1970, on the other hand, have begun to register their effects on native birth-rates.

All major linguistic groups of the native population have followed similar patterns of decline and upturn, but to different extents (Table 10). Most of the Eskimos escaped the initial adverse Russian impact and the smallpox epidemics of the nineteenth century, but the period of intense commercial whaling caused their numbers to drop by the turn of the century to about half their pre-contact levels. By the 1970 census they appear to have fully recovered. In contrast, the Aleuts suffered immediate and serious decline, falling to about 14 per cent of 1740 estimates by 1839

Table 10. Native population of Alaska by principal racial and linguistic groups, 1740–1970

Year[a]	Total	Eskimo	Aleut	Athapascan	Tlingit, Haida, Tsimshian	Not classified
1740	62,200	27,500	16,000	6,900	11,800	—
1839	39,000	24,500	2,200	4,000	8,300	—
1880	32,996	17,617	2,628	4,057	8,510	184
1890	25,354	13,871	1,679	3,520	5,463	821
1909	25,331	13,636	1,451	3,916	5,685	1,643
1920	26,558	13,698	2,942	4,657	5,261	—
1929	29,983	14,500[b]	4,500[b]	4,935	5,885	125
1939	34,458	15,576	5,599	4,671	6,179	433
1950	33,861	15,883	5,400[b]	4,700[b]	7,300[b]	600[b]
1970	51,128	27,985	6,581	8,937[b]	7,625[b]	—

[a] Data not available for 1900 and 1960.
[b] Allocation estimated from combined data.
(*Sources*: As Table 9.)

Table 11. Regional population trends in Alaska, 1740–1970

	Total Alaska	South-east	South-central	South-west	Interior	North-west
	(No of persons)					
Circa 1740–80						
Native	62,200	11,800	10,400	27,000	5,200	7,800
31 December 1909						
Native	25,331	5,866	3,205	7,326	2,403	6,531
Non-native	39,025	9,350	9,695	4,723	10,661	4,596
1 October 1939						
Native	32,458	6,502	3,974	10,858	3,322	7,802
Non-native	40,066	18,739	10,907	1,988	6,883	1,519
1 April 1970						
Native	51,128	8,248	9,005	17,296	5,751	10,828
Non-native	251,519	34,317	154,787	9,383	50,612	2,420
	(as per cent of State totals)					
Circa 1740–80						
Native	100.0	19.0	16.7	43.4	8.4	12.5
31 December 1909						
Native	100.0	23.2	12.7	28.8	9.5	25.8
Non-native	100.0	24.0	24.8	12.1	27.3	11.8
1 October 1939						
Native	100.0	20.0	12.2	33.5	10.2	24.1
Non-native	100.0	46.8	27.2	5.0	17.2	3.8
1 April 1970						
Native	100.0	16.1	17.6	33.9	11.2	21.2
Non-native	100.0	13.6	61.6	3.7	20.1	1.0

(*Sources*: As Table 9. Regions as defined in Map 15.)

and 9 per cent by 1909. Since 1920 they have increased steadily, but only to 41 per cent of the initial numbers by 1970. The effects of outside impacts on the Athapascan Indians were minimized, owing to their isolation and the scatter of their numbers over the vast interior region, and they had significantly exceeded their pre-contact levels by 1970. Smallpox and other diseases account for the early sharp decline in the numbers of the Tlingit and Haida Indians (the Tsimshians immigrated into Alaska from Canada in the 1880s). The failure to rise above the 1950 levels is due to increased migration out of Alaska attributable to shortening of commercial fishing seasons.

Changes in the regional distribution of non-native population can be attributed directly to stages in Alaska's economic development (Table 11). Although this distribution responded directly to change in location of economic activity, native population distribution appears to reflect only the biological factors inherent in the trends in Table 10. During the last decade there has been a significant increase in the participation of native people in the labour force, but for the past and to an important degree today they have remained in the subsistence economy of Alaska.

The historical record of changes in the location of population centres suggests the shifting patterns of inter-relations within Alaska. At the turn of the century the city of Nome, with a population of 12,488, was the largest community in Alaska, Juneau (1,864) was second, and Sitka (1,396) was third. Anchorage and Fairbanks did not exist, even as aboriginal populated places, and the rest of the population was scattered in isolated villages and settlements, most with fewer than 100 persons. There was no connection, except political, between Nome and the rest of Alaska. Nome declined but continued in first place until the 1910 census (2,600 persons), dropping far down the list in the following census. Ranking communities by size, Juneau remained the largest in the territory from the 1920 to the 1939 census with Ketchikan, Anchorage, and Fairbanks in second, third and fourth places respectively.

The extreme specialization of the several regional economies and the small size of the population concentrations made Alaska's communities heavily dependent upon centres outside the boundaries of the territory. The San Francisco Bay area was at first the centre providing capital, seasonal labour, entrepreneurial organization, and most of the supplies needed for economic development. The Puget Sound area progressively took over these functions until by the mid-1920s Alaska was for all practical purposes a colony of Seattle. Alaska's three largest towns functioned as trading and transport centres for a vaguely defined hinterland, but they were unable to provide many of the elements essential to the development process. The difficulties and lack of transport within Alaska resulted in only weak inter-relations between these towns and precluded the emergence of a single Alaskan centre with territory-wide influence.

Table 12. Regional distribution of native and non-native population of Alaska by type of place, 1939–70

Region and type of place	Native population				Non-native population			
	1 October 1939		1 April 1970		1 October 1939		1 April 1970	
	No. persons	%	No. persons	%	No. persons	%	No. persons	%
Southeast region	6,502	100.0	8,248	100.0	18,739	100.0	34,317	100.0
Urban[a]	1,863	28.7	4,868	59.0	13,333	71.2	28,909	84.2
Defence[b]	—	—	—	—	337	1.8	—	—
Rural	4,639	71.3	3,380	41.0	5,069	27.0	5,408	15.8
Southcentral region	3,974	100.0	9,005	100.0	10,907	100.0	154,787	100.0
Urban[a]	172	4.3	4,913	54.5	3,323	30.5	105,412	68.1
Defence[b]	—	—	340	3.8	—	—	26,801	17.3
Rural	3,802	95.7	3,752	41.7	7,584	69.5	22,574	14.6
Southwest region	10,858	100.0	17,296	100.0	1,988	100.0	9,383	100.0
Urban[a]	—	—	1,879	10.9	—	—	537	5.7
Defence[b]	—	—	54	0.3	—	—	5,686	60.6
Rural	10,858	100.0	15,363	88.8	1,988	100.0	3,160	33.7
Interior region	3,322	100.0	5,751	100.0	6,883	100.0	50,612	100.0
Urban[a]	100	3.0	1,358	23.6	4,343	63.1	29,260	57.8
Defence[b]	—	—	175	3.0	—	—	17,751	35.1
Rural	3,222	97.0	4,218	73.4	2,540	36.9	3,601	7.1
Northern region	7,802	100.0	10,828	100.0	1,519	100.0	2,420	100.0
Urban[a]	550	7.0	3,426	31.6	1,009	66.4	1,166	48.2
Defence[b]	—	—	—	—	—	—	178	7.4
Rural	7,252	93.0	7,402	68.4	510	33.6	1,076[c]	44.4

[a] *Urban places* here are taken to represent civilian communities of 2,000 or more population. 1970 boroughs used only where they embrace a natural community unit (core city and connected suburbs). 1939 place population for same geographic area embraced by 1970 data. Total urban population, by place:

	1939	1970
Southeast region		
Juneau borough	7,390	13,556
Ketchikan borough	5,631	10,041
Sitka borough	2,175	6,109
Petersburg city	(1,324)	2,042
Wrangell city	(1,162)	2,029
Southcentral region		
Anchorage borough	3,495	102,994
Kenai city	(303)	3,533
Kodiak city	(864)	3,798
Southwest region		
Bethel city	(376)	2,416
Interior region		
Fairbanks borough	4,443	30,618
Northern region		
Nome city	1,559	2,488
Barrow city	(363)	2,104

1939 population in parentheses ( ) not included in total 1939 urban population.
[b] *Defence places* include military and civilian population residing in listed defence reservations, bases and stations. 1939 listed only Ft Seward in southeast region as a military base.
[c] Includes petroleum workers at Prudhoe Bay and other North Slope sites.
(*Sources*: US Bureau of the Census Reports, PC(1)-A3, Alaska and PC(1)-B3, Alaska 1940 and 1970; University of Alaska, 'Age and sex characteristics of Alaska's population', *Review of Business and Economic Conditions*, March 1972; US Department of the Interior, 'Indian population, 1970 Census – Census of County divisions and places, Alaska', March 1971.)

Table 13. Selected social characteristics of Alaska population, by race, 1970

	White		Negro		Other non-white[a]	
	No. of persons	%	No. of persons	%	No. of persons	%
Age, sex, households						
Total population[b]	237,897	100.0	8,860	100.0	53,625	100.0
Male	130,466	54.8	5,188	58.6	27,475	51.2
Female	107,431	45.2	3,672	41.4	26,150	48.8
Both sexes under 18 yrs.	89,229	37.5	3,160	35.7	27,411	51.1
Both sexes 65 years and over	4,683	2.0	131	1.5	2,002	3.7
Median age	23.7		22.8		16.7	
In families	203,676	85.6	6,601	74.5	48,363	90.2
In other households[c]	15,554	6.5	592	6.7	3,359	6.3
In group quarters[d]						
military barracks	14,911	6.3	1,513	17.1	287	0.5
other group quarters	3,756	1.6	154	1.7	1,616	3.0
Mean size of families	3.60		3.78		5.78	
Residence five years ago						
Total persons, 5 years and older	231,828	100.0	7,762	100.0	46,699	100.0
Alaska	92,092	43.1	2,283	29.4	39,228	84.0
Outside Alaska	93,427	43.7	4,271	55.0	2,732	5.9
Not reported	28,308	13.2	1,208	15.6	4,739	10.1
Years of school completed						
Total persons, 25 years and older	110,906	100.0	3,597	100.0	20,445	100.0
Less than 5 years completed	1,109	1.0	176	4.9	6,677	32.7
Less than 1 year of high school	10,647	9.6	622	17.3	13,561	66.3
4 years of high school or more	83,512	75.3	2,320	64.5	4,178	20.4
4 years of college or more	18,299	16.5	281	7.8	448	2.2
Median school years completed	12.6		12.3		6.0	
Infant mortality						
(Deaths per 1,000 registered live births)	21.3		[e]		28.5	

[a] Alaskan natives represent 95.3 per cent of 'other non-white' races. These figures, therefore, include characteristics of 2,497 persons of Chinese, Japanese, Filipino, etc., ancestry.
[b] Data on characteristics not adjusted for census undercount.
[c] 'Other households' includes all persons occupying a single housing unit.
[d] 'Group quarters' include institutions (schools, hospitals, correctional institutions, etc.) and separate living quarters housing five or more unrelated persons (college dormitories, barracks, flophouses, etc.).
[e] Numbers too small to be statistically significant.
(*Sources*: US Bureau of the Census, Report PC(1)-C3 Alaska, Tables 48–52; Alaska Department of Health and Welfare, Vital Statistics.)

The shift of the basic economy from fishing and mining to defence and construction increased the population of Anchorage far ahead of all other communities. Defence activities not only advanced population at and around Anchorage and Fairbanks to the critical mass for achieving an important degree of independence from reliance upon outside centres, but also forged an improved transport and communications system. The more recent petroleum developments have broadened this base and jet aircraft have made travel-time and cost, not miles, the critical factor in setting new

Table 14. Estimates of total Alaska resident population and components of change, 1950–75 (thousands of persons)

1 July	Population composition[a] Total population	Military	Civilian	Components of annual change[b] Total	Natural increase	Net migration Military	Civilian
1950	138	26	112	8	2.5	(4)	9.5
1951	164	38	126	26	2.8	12	11.2
1952	196	50	146	32	3.8	12	16.2
1953	212	50	162	16	5.0	—	11.0
1954	218	49	169	6	6.1	(1)	0.9
1955	221	50	171	3	6.4	1	(4.4)
1956	220	45	175	(1)	6.5	(5)	(2.5)
1957	228	48	180	8	6.7	3	(1.7)
1958	213	35	178	(15)	6.5	(13)	(8.5)
1959	220	34	186	7	6.5	(1)	1.5
1960	228	33	195	8	6.3	(1)	2.7
1961	235	33	202	7	6.3	—	0.7
1962	243	33	210	8	6.4	—	1.6
1963	251	34	217	8	6.5	1	0.5
1964	256	35	221	5	6.5	1	(2.5)[c]
1965	267	33	234	11	6.3	(2)	6.7[c]
1966	272	32	241	5	5.8	(1)	0.2
1967	278	33	245	6	5.5	2	(1.5)
1968	285	33	252	7	5.4	—	1.6
1969	295	32	263	10	5.3	(1)	5.7[d]
1970	304.6	31.7	272.9	5.6	5.4	(0.3)	0.5
1971	312.9	30.1	282.8	8.3	5.5	(1.6)	4.4
1972	324.3	26.5	297.8	11.4	5.5	(3.6)	9.5
1973	330.4	27.5	302.9	6.1	5.5	1.0	(0.4)
1974	351.2	27.5	323.7	20.8	5.2	—	15.6[e]
1975	404.6	25.3	379.3	53.4	5.2	(2.2)	50.4[e]

[a] Estimates are 12-month moving averages centred on 1 July.
[b] Decreases and net out-migrations shown in parentheses.
[c] Reflects effects of 1964 earthquake and 1965 reconstruction.
[d] Reflects North Slope oil boom.
[e] Reflects trans-Alaska oil pipeline construction boom.
(*Sources*: 1950–1966 US Bureau of the Census, *Current Population Reports*, Series P-25; 1967–1975 Alaska Department of Labor, *Current Population Estimates, Alaska*.)

patterns of relations within Alaska. By the 1960s Anchorage-based banking and trade firms had expanded their business horizons beyond their immediate community to all parts of the State. Furthermore Anchorage, although located far from the North Slope petroleum developments, is the headquarters of these activities, and during the production phase will manage and control the process by rotating personnel and by electronic communications and control. The pattern is the same as that of defence activities with the Alaska command headquarters located at Anchorage. Alaska is still dependent upon outside centres for basic economic support, but Anchorage has taken over an increasing range of functions which formerly could be provided only at Seattle and elsewhere.

Table 12 summarizes the 1939 and 1970 distribution of native, non-native, and military population by five major regions and the local government units as they existed in 1970. The defence and urban centres in the southeast, southcentral, and interior regions account for 82.8 per cent of the total non-native population in 1970. The drift of native population to these dominantly non-native centres is indicated by the increase from 6.6 per cent of total native population in 1939 to 22.8 per cent in 1970. The rise of 'urban' centres within the southwest and northern regions reflects a regional concentration of surplus population no longer supportable by traditional village subsistence economies. These trends appear to be accelerating.

Table 13 presents selected characteristics of 1970 population. The male–female composition of native population has long been virtually in balance, but until recently the non-native sex distribution has been heavily weighted on the side of males. In 1890 and 1900 there were nine white males for every female, but by 1970 the ratio had dropped to 1.2:1, the remaining slight male excess being due to the large numbers of single men in the armed forces. Non-native age distributions reflect the practice of migration to Alaska during the active working years and leaving at retirement age. Only 43.1 per cent of white and 29.4 per cent of Negro population had lived in Alaska five years earlier, as compared with 84 per cent of 'other non-white' races (primarily native but including some orientals). The median years of school completed by this last category was less than half that of the whites and Negroes. The high proportion of native population under 18 years of age reflects their higher birth-rates. Infant mortality data are a good measure of the relative material well-being of natives and non-natives. The effectiveness of public health programs is reflected in the decline in native infant mortality rates from 95.9 per 1,000 live births in 1950 to 28.5 in 1970 (the white rate for the same period fell from 23.5 to 21.3).

Although net natural increase is again becoming an important component of change in total population, after dropping from its high levels in the mid-1950s to 1960s, net migration based upon changes in the resident economy continues to dominate (Table 14). Employment in commodity-producing industries and federal government, in fact, has been a major determinant of total population change. With appropriate leads and lags

Table 15. Annual changes in Department of Defense[a] and contract
construction employment and annual net migration,
Alaska, 1950–69
(thousands of persons)

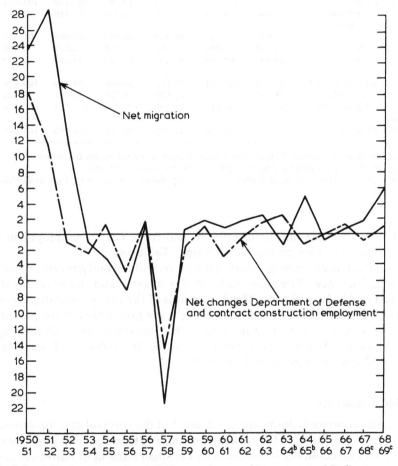

[a] Includes military personnel and civilian employees of Department of Defense.
[b] Distortion in normal migration due to 1964 Alaska earthquake and 1965 recovery (temporary evacuation and return).
[c] Reflects North Slope 'mining' activities in 1968–69.
Source: Alaska Department of Labor, *Annual Workforce Estimates* and *Estimated Current
Population*.

the annual net migration figures for the 1950s and 1960s, for example, have
almost equalled in absolute amount the annual net changes in defence and
construction employment with aberrations attributable to the great Alaska

Table 16. Geographical distribution of total Alaska population (July 1) 1970–5

	1970	1971	1972	1973	1974	1975
Total Alaska	*302,361*	*311,070*	*322,115*	*330,365*	*351,159*	*404,634*
North Slope and Upper Yukon[a]	9,099	9,354	9,136	9,320	11,119	23,657
Fairbanks[b]	50,043	48,147	49,837	49,856	55,266	61,411
Valdez-Chitina	3,098	2,392	3,464	3,568	3,833	9,639
Total pipeline corridor	*62,240*	*60,463*	*62,437*	*62,744*	*70,218*	*94,707*
Anchorage-Matanuska	132,833	142,264	151,565	158,026	162,899	190,579
Kenai-Cook Inlet	14,250	14,204	13,830	13,808	13,962	15,621
Total other petroleum areas	*137,083*	*156,468*	*165,395*	*171,834*	*176,861*	*206,200*
Balance of State	93,038	94,139	94,283	95,787	104,080	103,727

[a] Includes Barrow, Upper Yukon, and Yukon-Koyukuk 1970 Census Divisions.
[b] Includes Fairbanks and Southeast Fairbanks 1970 Census Divisions.
(*Source*: Alaska Department of Labor, *Current Population Estimates by Census Divisions* (annual).)

earthquake of 1964 and its aftermath, and from 1967 to employment in oil and gas activities on the North Slope (Table 15).

A similar pattern emerges during the 1970s with oil and gas construction replacing defence. The magnitude of this latest boom, however, is the greatest in Alaska's history. The dimensions of the 1970s population boom are highly localized. Betwen 1970 and 1975 the population of the pipeline corridor area increased 52.2 per cent, other petroleum areas (Anchorage and Kenai-Cook Inlet) 50.4 per cent, while the balance of the State increased only 11.5 per cent (Table 16).

Resident economy

The cash economy is dominant in Alaska, but the survival of a subsistence economy is a feature of the rural native areas. The extent and importance of this second economy is indicated by the small number of natives in paid jobs, their relatively low *per capita* cash income, and the importance of fish and game as food. The cash economy is very simple. The basic sector is generated by federal spending (primarily defence) and extraction, harvesting, and processing of petroleum, pulp, timber, and fisheries products. There are other, less important, basic industries and in the past the major ones were different (e.g. canned salmon, gold, and copper before the Second World War). Most of the key economic decisions relative to each of these industries are made outside Alaska, the important economic decisions within Alaska relating to devising means of maximizing the benefits to the resident economy and providing the secondary or support sectors of the economy.

Statistics on net value of products from Alaska's natural resources, expenditures on defence and supporting activities (Gross State Product, Table 22), and foreign and domestic imports and exports (Table 23) indicate the total national and international economic importance of Alaska, but can be misleading as a representation of the resident economy. Estimates of personal income received by residents of Alaska, available annually since 1950, give a better description of the structure and trends of the Alaskan economy, but over time are subject to the distortions of monetary and wage inflation. Annual data on the employed workforce by economic components and industrial classification are the best indication of shifts in the structure of the Alaskan economy over time (Tables 17 and 18).

Value of commodity production rose almost six-fold between 1950 and 1970, but because of the capital-intensive nature of much of this activity the rise in employment in commodity-producing industries was modest. The 1950 level was not exceeded until 1967, with subsequent significant rises coming primarily from the oil and gas and the construction industries.

Support industries (transport, communication, public utilities, wholesale and retail trade, finance, insurance, real estate, and services) and non-defence government employment grew dynamically, causing continual change in the structure of the total economy from 1959 to 1969. Part of this expansion was a 'catching up' of the support component of the economy with developments that had already taken place in the more basic components (in Alaska's case the defence build-up of the earlier 1950s). It also reflects shifts in communications functions from the military to the civilian sector, growth of tourism, increased urbanization, and shift of headquarters and wholesaling functions from Seattle to Anchorage.

Federal government employment in Alaska has always been abnormally high relative to total employment for several reasons. These include the special programs associated with managing the huge acreage of public domain lands (46.4 per cent of all lands of the United States managed by the federal government as of 30 June 1970 were located in Alaska), the needs of the native peoples who were nominally wards of the federal government (approximately one-fifth of Alaska's total population), and the basic transport and communications functions performed by federal agencies that are private responsibilities elsewhere in the United States. Increases in state and local government employment reflect transfer of certain functions from the federal government and greater revenues from federal grants, oil and other natural resource royalties and lease bonuses, and income tax.

Table 17. Industrial composition of total Alaska labour force, 1939–75

Industrial classification	Calendar years (twelve month averages. thousands of persons)															
	1939	1950	1952	1958	1960	1965	1966	1967	1968	1969	1970	1971	1972	1973	1974	1975
Total employed	26.6	78.1	111.6	96.0	100.7	115.2	117.5	121.7	123.7	128.7	130.2	133.9	136.5	143.2	161.5	192.9
Department of Defense[a]	0.6	32.0	57.5	42.5	39.5	39.5	39.7	40.4	39.4	39.0	38.7	36.0	32.3	32.8	33.1	30.7
Commodity-producing industries[b]																
Agriculture	1.0	1.0	0.9	0.9	0.8	0.8	0.8	0.8	0.8	0.8	0.8	0.8	0.8	0.8	0.8	0.8
Fishing	1.5	1.5	2.0	1.8	1.8	2.6	2.7	2.7	3.2	2.8	3.1	2.8	2.9	3.1	3.0	3.0
Mining																
Crude oil and gas	—	0.5	0.4	2.0	0.4	0.7	1.0	1.6	2.2	3.2	2.7	2.1	1.8	1.7	2.6	3.4
Other	4.5	1.4	1.3	0.9	0.7	0.4	0.4	0.4	0.3	0.3	0.3	0.3	0.3	0.3	0.4	0.4
Contract construction	1.3	6.3	10.3	5.1	5.9	6.4	5.9	6.0	6.0	6.7	6.9	7.4	7.9	7.8	14.1	25.9
Manufacturing																
Food processing	4.2	4.7	4.7	3.0	2.8	3.0	3.4	3.1	3.3	3.2	3.7	3.6	3.8	4.6	4.3	4.3
Logging, lumber, pulp	0.1	0.6	0.8	1.1	2.2	2.3	2.3	2.6	2.5	2.5	2.8	2.8	2.8	3.2	3.6	3.4
Other (including petroleum products)	0.3	0.4	0.5	0.8	2.1	1.0	0.9	0.9	1.1	1.4	1.3	1.4	1.5	1.5	1.7	1.9
Support industries																
Transportation communications, utilities																
Air transportation	0.5	0.9	1.4	1.6	2.0	1.9	2.0	2.2	2.5	3.1	2.8	2.8	3.0	3.3	4.0	4.8
Other	1.5	2.8	3.1	4.1	4.8	5.3	5.3	5.3	5.3	5.7	6.1	7.0	7.0	7.1	8.4	11.7
Trade	2.5	4.9	6.7	6.6	7.7	10.0	10.8	11.8	12.5	14.0	15.4	16.2	17.1	18.3	21.1	26.2
Finance, insurance, real estate, services & miscellaneous	1.4	4.6	6.1	8.1	9.1	12.8	13.2	14.1	14.8	14.4	14.5	15.8	17.7	19.5	23.2	32.1
Government (non-defence)																
Federal (non-defence)	2.0	7.6	7.0	9.3	9.2	9.8	9.8	10.0	10.1	9.9	9.8	11.3	11.4	11.8	12.4	12.9
State (Territorial), local	1.0	2.1	2.5	5.0	7.1	13.4	14.4	14.4	15.3	16.8	18.4	20.7	23.3	24.5	25.8	28.9
Unclassified[c]	4.2	6.8	6.4	5.0	6.0	5.3	4.7	4.7	4.4	5.0	2.7	2.9	2.9	2.9	2.6	2.5

[a] Includes military personnel and civilian employees of the Department of Defense (Departments of War and Navy in 1939).
[b] Includes agricultural wage and salary and unpaid family workers.
[c] Unclassified self-employed, unpaid family workers, etc. Basis of estimation changed after 1969.

(Sources: 1939: Estimated from 1939 census and 1940 covered employment data, 1950–1969: Alaska Department of Labor, *Workforce Estimates, by Industry* (Annual); 1970–74: Alaska Department of Labor, *Labor Force Estimates, by Industry and Area* (basis changed to eliminate multiple job holdings); Military personnel from Alaska Department of Labor, *Current Population Estimates* (Annual). Fishing estimates based on George W. Rogers and Richard Listowski, 'A study of the socio-economic impact of changes in the harvesting labor force in the Alaska salmon industry', National Marine Fisheries Service, US Department of Commerce, December 1972; 1971–76 estimated on basis of change in licences.)

Table 18. Relative changes in industrial composition of total Alaska employed workforce, 1939, 1952, 1972

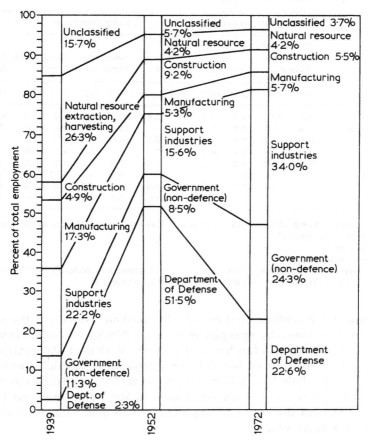

The geographical distribution of the economic impact of the trans-Alaska pipeline construction is shown in the statistics on employment by place of residence of the worker (labour force) in Table 19. Between 1970 and 1975 the number of employed workers living in the pipeline corridor (including those in construction camps) are increased by 90 per cent. The headquarters of the project and of other oil and gas development activities and the location of the home base of many of the workers is the Anchorage area, which experienced a 41 per cent increase in employed labour force. In contrast, employment in the balance of the State increased only 12 per cent.

For the rest of the 1970s construction of facilities related to gas and oil

Table 19. Alaska employment by petroleum-related and other areas, 1970 and 1975 (thousands of persons, monthly averages)

	Trans-Alaska pipeline corridor						Other petroleum					
	North Slope and Upper Yukon[a]		Fairbanks		Valdez-Chitina		Anchorage Matanuska[a]		Kenai-Cook Inlet		Balance of State	
	1970	1975	1970	1975	1970	1975	1970	1975	1970	1975	1970	1975
Total Employment	4.5	12.9	23.2	37.2	1.1	4.6	60.2	85.9	4.9	5.7	36.1	44.1
Military personnel	0.9	0.9	8.8	6.1	—	—	12.9	12.6	0.5	0.1	8.3	5.6
Government	0.8	1.8	6.5	7.5	0.5	0.7	16.0	21.0	0.8	1.1	11.2	15.2
Mining	1.0	1.2	0.1	0.2	[b]	[b]	1.0	1.3	0.7	0.8	0.2	0.3
Construction	0.5	7.3	1.2	7.4	[b]	2.5	3.6	7.2	0.4	0.6	1.2	0.8
Manufacturing	—	—	0.2	0.5	[b]	[b]	1.0	1.6	0.6	0.9	5.9	6.7
Support industries	1.2	1.6	6.5	14.9	0.3	1.4	21.4	40.6	1.3	2.3	9.0	14.1
Other[c]	—	—	—	0.7	0.2	—	4.2	1.6	0.8	—	0.2	1.3

[a] Includes 1975 Barrow–North Slope, Upper Yukon and Yukon–Koyukuk labour market Divisions and 1970 Census geographic equivalents.
[b] Less than 50 persons.
[c] Unclassified self-employed, etc.
(*Source*: Alaska Department of Labor, *Alaska Labor Force Estimates by Industry and Area, 1970, 1971, 1972*, revised July 1975. Preliminary 1975 estimates subject to revision.)

development will continue to dominate the Alaskan economy. Unless the oil pipeline is followed by a gas pipeline through Alaska, petroleum-related employment will fall to a few hundred workers. With the full operation of the pipeline, production of North Slope petroleum is scheduled to be at 2 million barrels daily (100 million tonnes a year) for the next two to three decades. Large revenues will be generated for both state and local governments, and the resulting employment will dominate the Alaskan economy for many years.

Labour force

Table 20 summarizes several aspects of the Alaskan labour force as compared with the United States in 1965 and 1972. More recent comparisons are not made because of the abnormalities introduced by the trans-Alaska pipeline construction boom. The figures for 1972 are probably more representative of a future sustainable pattern. In both years the Alaska population had a relatively high participation rate in the labour force compared with the United States as a whole, despite the extremely low participation rates among rural native Alaskans (Table 21). This reflects unusual age-sex profiles of Alaska's non-native population, the high number of single males in the armed forces and construction and

Table 20. Comparison of major characteristics of Alaska and United States labour force, 1965–72 (thousands of persons)

	Alaska		United States	
	1965	*1972*	*1965*	*1972*
Total population – labour force[a] *relations*				
Total population, July 1	265.2	324.3	194,303	208,700
Total labour force (including armed forces)	122.8	157.7	77,178	88,991
Dependency ratio (persons not in labour force divided by total labour force)	1.16	1.06	1.52	1.35
Total civilian labour force	89.8	131.2	74,455	86,542
Total unemployed	7.7	13.6	3,366	4,840
Per cent civilian labour force	8.6	10.4	4.5	5.6
Industrial composition, annual average	(per cent)			
employed labour force				
Government (including members of armed forces)	*54.5*	*46.5*	*17.3*	*18.7*
Private employment	*45.5*	*53.5*	*82.7*	*81.3*
Agriculture	0.7	0.6	5.9	4.1
Non-wage, non-agricultural	9.4	8.7	8.0	6.5
Wage and salary support[b]	23.4	31.6	39.1	43.3
Commodity-producing	12.0	12.6	29.7	27.4
(Mining)	(1.0)	(1.5)	(0.9)	(0.7)
(Contract construction)	(5.6)	(5.5)	(4.3)	(4.2)
(Manufacturing)	(5.4)	(5.6)	(24.5)	(22.5)

[a] Labour force, employment, and unemployment data represent annual averages.
[b] 'Support' industries include transport, communications, utilities, trade, services, finance, insurance, real estate.
(*Sources*: US Department of Labor, *Manpower Report of the President*, April 1975; Alaska Department of Labor, *Alaska Labor Force Estimates by Industry and Area* (annual); Alaska data not adjusted for fishing estimates as done in Table 17, above.)

petroleum development crews, and the large number of wives employed. Also apparent is the dominant place of government employment in the industrial composition of the annual average employed force (Table 17) and an average annual civilian labour force unemployment rate almost twice the national average. Much of the difference can be accounted for by the fact that a substantial proportion of the Alaska 'unemployed' at any particular time are not permanent members of the State's labour force (i.e. non-resident summer migrants, seasonally employed house-wives, etc.). Moreover, seasonal factors account for a very large portion of total unemployment; in the calendar month of lowest unemployment in Alaska the State unemployment rate is typically only about two percentage points above the national figure. On the other hand, the difference between Alaskan and US unemployment rates would be even greater if all rural joblessness were counted as unemployment in official State reports. Lack of work among rural natives, for example, now largely appears only as low labour force participation rates.

Table 21. Selected economic characteristics of Alaska population, by race, 1970

	White		Negro		Other non-white[a]	
	No. of persons	%	No. of persons	%	No. of persons	%
Employment status, April 1970, by sex						
Males, 16 years and older	88,945	100.0	3,789	100.0	14,543	100.0
In labour force[b]	80,072	90.0	3,554	93.8	7,397	50.9
Armed Forces	27,582	31.0	2,573	67.9	575	4.0
Civilian labour force	52,490	59.0	981	25.9	6,822	46.9
Employed	47,931	53.9	902	23.8	5,281	36.3
Unemployed	4,559	5.1	79	2.1	1,541	10.6
Females, 16 years and older	67,631	100.0	2,092	100.0	14,127	100.0
In labour force[b]	33,097	48.9	1,216	58.1	4,385	31.0
Armed forces	627	0.9	58	2.8	10	0.1
Civilian labour force	32,470	48.0	1,158	55.3	4,375	30.9
Employed	30,300	44.8	1,057	50.5	3,765	26.7
Unemployed	2,170	3.2	101	4.8	610	4.3
Labour mobility of males, 30–49 years old in 1970[c]						
Males, 30–49 years old	34,756	100.0	1,235	100.0	5,421	100.0
Non-worker in 1965, non-worker in 1970	744	2.1	28	2.3	1,481	27.3
Non-worker in 1965, worker in 1970	2,736	7.9	92	7.4	748	13.8
Worker in 1965, non-worker in 1970	2,215	6.4	32	2.6	1,102	20.3
Worker in 1965, worker in 1970	29,061	83.6	1,083	87.7	2,090	38.6
Income of families, 1969						
Median income	$13,464		$8,638		$5,092	
Mean income	14,874		9,917		7,801	
Mean income per family member	4,130		2,625		1,418	
Percentage of families receiving public assistant income (included in total income, above)	2.0%		4.8%		22.9%	
Income of unrelated individuals, 1969						
Median income	$ 3,706		$2,671		$1,585	
Mean income	5,712		3,889		2,906	

[a] Alaska natives represent 95.3 per cent of total 'other non-white' races. These data include characteristics of a minority of persons of Chinese, Japanese, Filipino, etc., ancestry.
[b] 'In labour force' means working or actively seeking employment.
[c] 'Worker' includes members of the armed forces.
(*Source*: US Bureau of the Census Report PC(1)-C3, Alaska, Tables 53–8.)

A comparison of the levels of monthly employment in selected industrial categories indicates the magnitude of seasonality as a characteristic of the Alaskan economy. In 1970, the total employment level in July was about one-third above the level of January. Employment was relatively stable seasonally in government and support industries. In commodity-producing industries the abnormally high seasonal fluctuations ranged from a ratio of the peak month to low month of 20 to 1 in commercial fishing, 5 to 1 in fish processing and 1.5 to 1 in other manufacturing.

Alaska is notorious for stories of high rates of pay and income. There is some basis in fact for these impressions, but they are not offset by the instability of employment seasonally and secularly. Average hourly earnings in contract construction in Alaska in 1975 were estimated at $15.05 as compared with the national average of $7.25. In retail trades, the Alaskan average hourly earnings were $6.40 as compared with $3.34 nationally in 1975. This in part reflects higher wage-rates in Alaska, which are justified by higher costs of living and in some cases remoteness of location, but earnings also reflect overtime payments. In contract construction in Alaska in 1975, the average hours worked per week were 55.3 hours (in October 65.7 hours) as compared with 36.6 hours per week nationally.

Statistics on native Alaskans' employment and income status are not available from census reports, but natives are the dominant racial group in the 'other non-white' category and this can be taken as a proxy. Table 21 compares white, negro, and other non-white population and income characteristics. Not only do these data indicate the low level of participation of natives in the labour force (50.9 per cent of males 16 years and over, for example, as compared with 90 per cent of white males 16 years and over), but their rates of unemployment are about twice that of whites and Negroes. Their labour mobility (changes from non-worker to worker status) is also markedly lower and their family incomes about half that for white families.

Undoubtedly there is some institutional discrimination against natives in employment, and the difficulties of making the transition from the freedom of traditional life to the discipline of wage employment may make many natives 'unemployable'. The most important measurable influences on native unemployment, manifest and disguised, are their location outside the principal areas of economic activity (Table 12) and their low level of formal education (Table 13).

Government and administration

Political development

The course of economic development in Alaska has been influenced by political changes produced in turn by conflicts of economic interest and differential effects of this development upon social groups and public welfare. The purely economic forces of simple colonial exploitation must now accommodate themselves to the changes in the political environment wrought by the creation of Alaska as a 'sovereign' State of the Union in 1959, the emergence in the 1960s of the native Alaska population as an effective political force, and the native land claims settlement of 1971.

From 1867 to 1884 there was no civil government in Alaska beyond its creation as a United States Customs District under the jurisdiction of district courts in Oregon or California. The gold stampedes brought

popular demands for elaboration of government and administration. Under
the first Alaska Organic Act (17 May 1884) Alaska became a civil district,
a judicial district, and a land district. A governor was appointed by the
President of the United States to administer the provisions of the Act.
Homestead laws were extended to Alaska in 1898. The Civil Code of 1900
granted Alaskans the authority to establish local municipalities, and in 1906
the right to elect a voteless delegate to Congress. The second Organic Act
(24 August 1912) completed the creation of Alaska as a territory of the
United States with its own popularly elected Legislature. But the Governor
was still a Presidential appointee, the actions of the legislature were
subject to veto by the US Congress and the courts, and most basic
governmental functions remained beyond local control or influence.

During this period economic development followed traditional colonial
lines in ignoring local interests and in being specialized and exploitative.
National conservation movements and the public scandals of the opening
decades of the present century also gave a strong preservationist bent to
later federal land policy, causing further frustration of plans developed
locally.

In 1958, the last year of Alaska's territorial status, 99.6 per cent of the
land area and its resources were in federal ownership and control, and
fisheries resources were still under federal management.

This combination of private and public absentee control thwarted
resident aspirations and created popular demand for greater self-
determination. The conflict between resident and non-resident interests
was transformed into an effective grass-roots political movement by the
catalyst of population growth associated with defence development, and
Alaska became a State in 1959. This shifted the balance of ownership or
control of land and resources, and by decentralizing political power
included resident as well as non-resident interests in shaping the objectives
of economic development.

The creation of the State of Alaska on 3 January 1959 was a ceremonial
event of the highest political order. In effect, the people of the United
States acting through their duly elected representatives, the members of
the Congress and the President of the United States, entered into a
compact (so described in the Act) with the people of Alaska, acting
through a popular referendum, by which sovereign powers and responsibilities were shared and certain institutional rearrangements and conditions agreed upon. Alaskans were granted voting representation in the
US Congress (two Senators and one Representative), and the right to
participate in the election of the President of the United States and to
establish their own institutions and systems of government within their
boundaries, providing the form was republican and in no way 'repugnant
to the Constitution of the United States and the principles of the Declara-

tion of Independence'. The Act also offered the new State the right to select 103.33 million acres (410,000 sq km) from the public domain (27 per cent of the total land area of the State) and, immediately following statehood, the Secretary of the Interior transferred all responsibility for the management of commercial fisheries resources to the new State.

Natives had a low level of participation in development and its economic and social benefits, and suffered loss of lands and resources essential to maintaining their traditional ways of life. Organized protest was weak, sporadic, and ethnically and geographically divided. This changed in the mid-1960s when concern arising from a combination of major economic and public works proposals brought into being regional protective associations which then united under the Alaska Federation of Natives, the first effective statewide Eskimo, Aleut, and Indian political movement. Formal claims were made to the title to all the lands of Alaska on the basis of aboriginal use and occupancy, and with the support of the petroleum industry and government of the State of Alaska, the Federation secured passage by the US Congress in December 1971 of the Alaska Native Claims Settlement Act (Public Law 92–203, 92nd Congress). In return for extinguishing aboriginal claims, the native people will receive title to 44 million acres (176,000 sq km) of land and the mineral estate, grants totalling $462.5 million from the Federal Treasury payable over eleven years, and $500 million from 2 per cent of the annual revenues from mineral leasing activity on state and federal lands.

The Act also organizes the native people into a system of interlocking regional and village corporations for the purpose of promoting their fullest economic and social development. The geographical boundaries of the twelve regional corporations approximate to the 'territories' of the twelve defined major ethnic groups of Alaska's aboriginal population. Taken as a whole, the Act is as much a development as a settlement Act. In a new departure from the traditional approach to dealing with minority peoples, native Alaskans have the opportunity of working out their own destinies with a generous endowment of land, money, and organization.

Just as Alaska's turn-of-the-century development coincided with the flowering of the national conservation movement, the Alaskan developments of the 1960s coincided with national concern over the crisis in outdoor recreation and, on the threshold of the 1970s, the crisis of the environment. Private and public developers must now meet the requirements of the National Environmental Policy Act of 1969, as well as counter direct intervention by organized national conservation and environmentally concerned groups. These political forces, as a trade-off for not opposing the legislation, were also able to get an environmental and conservation rider attached to the Native Claims Settlement Act of 1971 which had little to do with its main purpose. This provided for expansion of the nation's parks, forests, wildlife refuges, and wild and scenic river systems through permanent reservation of 80 million acres (320,000 sq km) of presently unoccupied and unreserved public domain lands in Alaska.

Structure

The Constitution of the State of Alaska presents a formal statement of the institutional arrangements by which Alaskans agree to govern themselves. Executive power resides with the Governor, perhaps more forcefully than it does in any other State. He and the Secretary of State are the only elected executive officials. He has the usual duties of enforcing the law and the power to veto legislative actions. He appoints the head of each executive department, appoints members to regulatory and quasi-judicial boards subject to legislative confirmation, and has a key role in filling vacancies in the court system.

The legislature is bicameral, composed of a Senate of twenty members elected for four-year terms from eleven districts and a House of Representatives of forty members elected for two-year terms from nineteen districts. Regular sessions convene annually in January and are not limited in length. The Governor may call special sessions. The Alaskan court system is the third branch of state government. At the bottom are fifty-two magistrates followed in the hierarchy by seventeen district judges and eleven superior court judges. The five-man Supreme Court has final appellate jurisdiction.

There are two forms of local government, boroughs and cities. Boroughs are an intermediate level of government corresponding approximately to counties in other States. The entire State is divided into organized boroughs or left as an unorganized borough. Locally elected asemblies are the governing bodies of organized boroughs, and the State legislature governs the unorganized borough. Organized boroughs are established by local populations, but the areas served are based on natural economic units, which include the surrounding natural resource base. There are three classes of organized borough and four classes of city, defined by functions to be performed and population size. An organized borough and all cities included within it may unite to form a single unit of home-rule local government called a unified municipality.

The federal government continues to play an important role in Alaska's government and economy. Total federal employment (including military personnel) accounted for 37 per cent of all employment in 1970, and for 23 per cent in 1975, at the height of the pipeline construction boom. The Department of Defense has the greatest impact upon the resident economy, but other departments and agencies perform vital functions in managing the extensive public domain lands, providing or subsidizing transport, and providing the usual range of federal social and other services. After all land transfers have been made to the State and to the native corporations (circa 1984), the federal government will still own and manage 57.7 per cent of Alaska's lands.

A number of institutional arrangements contained in the Alaska Native Claims Settlement Act of 1971 have important implications for government and administration. The use, development, and control over the lands and money granted to the native people under the Act will be through twelve regional corporations within Alaska, a thirteenth corporation formed for natives who are not residents of the State, and 223 village corporations. The regional corporations are established under state law as business-for-profit corporations and the village corporations may be for profit or non-profit. Citizens of the United States of at least one-fourth degree of Alaskan Indian, Eskimo, or Aleut blood born before 18 December 1971 became members and shareholders of the applicable corporations with a voice in management and a share in the lands, assets, and income of the corporations.

Transport

Transport systems reflect the region's economic development, the requirements of the population served, geography, and the technology available (see Map 11). A realistic formula also includes the inertia of the conventional wisdom, the impacts of historical accident, the irrationality of popular preference as to modes of transport, and the politics of public spending. Logically, therefore, the treatment of transport systems follows presentation of these aspects of Alaska, which called forth and condition their development.

For transport purposes, Alaska, on the brink of the Second World War, was an island that could be reached only by sea. Passengers and freight were moved either by Canadian ships operating between Vancouver and Skagway or by the motley fleet of American relics salvaged for the most part from other maritime trades. There were no land links with the rest of the continent and the only air link was provided by a flying boat operated by Pan American Airways between Seattle, Ketchikan, and Juneau as a first pioneering step towards a trans-Pacific route. Bush pilots provided air communications elsewhere, landing on water or gravel strips.

To serve a land mass of 586,400 square miles (1,520,000 sq km) there were only 2,500 miles (4,000 km) of dirt and gravel road and trails. The Alaska Railroad, operating since 1923 but never quite completed, hauled passengers and freight along a single track 470 miles (770 km) from Seward to Fairbanks, using coal-burning locomotives and wood-frame cars salvaged from the Panama Canal construction. The journey took two days with an overnight stop mid-way at Curry Hotel. The White Pass and Yukon Route, built during the Klondike gold rush, connected the port of Skagway with Whitehorse at the head of the river transport in the Yukon Territory. The Copper River and Northwest Railroad served copper mines between 1911 and 1938.

Except for railroads, movements within Alaska were very much as they had been before the coming of the Europeans and Americans – by water through the Inside Passage of southeastern Alaska, along the coast, and on the major river systems of the interior, and in winter over a wider area by dog team. These primitive systems served the needs of the colonial economy of pre-war Alaska.

Reviewing Alaska's transport needs in 1937, the National Resources Committee considered the sparse and scattered population, limited and shifting economic activities, and the formidable land barriers. It recommended against heavy capital investment in fixed land routes until population-development patterns were established, and favoured development of air, river, and coastal marine facilities capable of overcoming or going around topographical obstacles: in short, evolving a set of special northern transport systems rather than transplanting more southern varieties. War and political factors, however, decreed otherwise.

The war in Alaska opened in June 1942 with the occupation by Japanese forces of Attu and attacks on Dutch Harbor followed by the fifteen-month 'thousand mile war' required to regain complete United States control of this Alaskan frontier. The conduct of this campaign and the subsequent build-up of neglected defences in Alaska were hampered by lack of adequate sea transport, hazards of sea and weather, and fear of enemy submarines. Alaska could no longer be treated as an island. Military necessity required that the basic transport system exploit Alaska's attachment to the North American continent. Accordingly, the Alaska Highway was pushed through the Canadian wilderness and the primitive road system within Alaska was hurriedly laced together and extended. By 30 June 1958, the last fiscal year before statehood, the pre-war mileage of roads and trails of all grades more than doubled to 5,196 miles (8,350 km), the main system was completely paved, and the Alaska Highway became an all-year continental link.

All of this was done without regard to economic considerations, military necessity overriding any question of cost. But the result was a major impact upon all transport modes through economies of scale. More directly, military needs provided surfaced air strips and fields and commercial aviation leapt forward with the use of more advanced aircraft. The number of civilian aircraft registered increased from 101 in 1937 to 1,208 in 1959. Revenue passengers increased from 20,958 in 1936 to 534,700 at international airports and 560,630 at all other airports in 1962, and air freight from 1 ton in 1936 to 61,000 tons in 1962. In 1948, through defence justification, the Alaska Railroad was completely rebuilt and modernized, although the line was not extended. Freight tons hauled increased from about 150,000 in 1937 to over 1,250,000 in 1959. Of this amount just over half was military cargo and much of the remainder was construction and other military-related cargo. In 1956 the Haines to Fairbanks petroleum products pipeline, built by the army, went into operation.

Waterborne commerce between Alaskan and other United States ports increased from 873,600 short tons (790,000 tonnes) of dry cargo and petroleum products in 1936 to 4.7 million short tons (4.25 million tonnes) in 1959. Port facilities at Seward and Valdez were improved by the federal government and a new military port created at Whittier. This quantum jump in scale of operation encouraged such major innovations as container-ships, hydro-trains (rail-water-rail links), ocean-going tugs and barges, and other vessels specifically designed to meet Alaskan conditions and needs. The savings in freight rates by these new marine systems has virtually eliminated the use of general cargo ships. In February 1971, Alaska Steam, which had dominated the maritime trade until the 1950s, announced termination of its cargo services.

Since Alaska became a State in 1959, transport systems brought into being by military necessity have been changed and expanded to serve the burgeoning civilian population better. The new Alaskans settled for the most part in specialized occupations in the two major centres, Anchorage and Fairbanks. Highway construction was transferred to State control, thus shifting the emphasis from long-term economic development to meeting the amenities desired by the new urban and car-owning electorate.

More importantly, Alaska was included in the federal-aid highway program commencing in 1956. The federal share of highway construction costs under this program in Alaska is 95 per cent because of the high percentage of land area still under federal control. The program also requires construction standards exceeding pioneering needs. Generous federal funding and standards made road building an end in itself (support of the construction industry and employment) rather than a means to the end of development. Growth in road mileage was modest, but constant re-building and upgrading of existing routes increased the speed, safety, and convenience of their use. Total route mileage within the Alaska State Highway System increased from 4,249 miles (6,830 km) by 30 June 1963 (the date at which the system had been brought substantially up to standard) to 4,900 miles (7,900 km) ten years later.

Major transport development often owed little to economic or even rational considerations, but automatically reflected the provisions of the federal matching-funding formula. Roads were the most extreme example of this political impact, but a similar pattern appeared in other modes. Airborne commerce is by private carriers, but they must operate within a legal and operational framework of federal and state governmental policies and programs. The Federal Aviation Agency provides major operating subsidies to carriers, maintenance of air navigational aids, and matching grants to the State and local governments for construction and maintenance of airports and facilities. The federal Civil Aeronautics Board certifies the routes of inter-State and intra-State carriers, determines tariffs, mail service, and amounts of subsidies, and investigates and reports causes of accidents. Using federal and state funds, the Department of Public Works plans, constructs, operates, and maintains the state airport system consisting of the Anchorage and Fairbanks international airports and about 400 other facilities (trunk and secondary airports, recreational flying strips, and float plane facilities). Airline service ranges from scheduled jet flights several times daily between trunk airports and some secondary airports to weekly flights of single-engined aircraft to bush communities.

The Alaska Marine Highway system, inaugurated in 1963 and financed by traffic revenues and state funds, includes 2,300 miles (3,700 km) of water routes within Alaska, connecting twenty-five coastal communities having a combined 1970 population of 53,000. Beyond the State it is connected to Seattle by weekly service, and Prince Rupert in British Columbia more frequently. The system serves the northern Gulf of Alaska, Prince William Sound, Cook Inlet, and Kodiak Island with two ocean-going car ferries, and southeastern Alaska with five large vessels on the main line and four smaller vessels serving other communities. Long-range plans include connecting these two parts of the system, and its extension into the Bering Sea and up the main river systems of the interior. In a sense this represents an attempt to re-establish the original transport systems with up-dated technology. The system has also been very popular with tourists, and in combination with private trucking provides the basic freight movement for the southeastern region.

The major natural resource developments of the 1950s and 1960s were aided by these basic transport systems, but their existence was not necessarily a prerequisite. In a straight line the North Slope discovery wells at Prudhoe Bay, for example, are on the far north side of the Brooks Range some 290 miles (470 km) from the nearest spur of the interconnected road system and 340 miles (550 km) from the head of the Alaska Railroad at Fairbanks. Throughout the exploratory period supplies and equipment were flown around the clock to drilling sites by fleets of C-130 Hercules transports carrying loads of 22 short tons (20 tonnes) and capable of landing on quickly made air strips. Movement by air increased freight handled by the Fairbanks airport, the gateway to the North Slope, to

120,300 short tons (110,000 tonnes) in 1969 (six times the highest amount handled during any previous year) and 180,202 short tons (165,000 tonnes) in 1975. Rigs were transported overland by cat-train, down Canada's Mackenzie River and into the Arctic Ocean, or in barges hauled by tugs some 2,500 miles (4,000 km) through the Bering Sea and into the Arctic Ocean.

During the field-development period in 1970 waterborne freight to the North Slope was shipped by convoys of tugs and barges. More than seventy ocean-going vessels carried 185,000 short tons (160,000 tonnes) of supplies and construction materials, including 169 miles (272 km) of 48-inch (122 cm) diameter pipe, through the Bering Strait to Prudhoe Bay. Two 100-ft by 400-ft (30×120 m) pipe barges, each towed by a 7,000 hp tug, hauled cargo between Tacoma, Washington, and Alaska and four directly from Japan to the North Slope. Quad-track vehicles, capable of moving heavy drilling rigs and freight weighing up to 30 tons across swampy muskeg country and tundra during summer, eliminated costly delays of waiting for winter freeze-up or construction of roads which would have cost millions of dollars.

A variety of methods of getting the product to market when production commences, were considered. In 1969–70 the tanker *Manhattan* (115,000 dwt), fitted with an icebreaker bow, demonstrated the physical feasibility of surface transport via the northwest passage, but the project was not followed up, presumably because of economic factors. Early in 1969 Atlantic Richfield, British Petroleum, and Humble Oil and Refining made a joint announcement of intention to build a 48-inch (122 cm) pipeline to move oil from the North Slope 800 miles (1,300 km) southward to Valdez, a year-round port on the Gulf of Alaska, at the rate of 2 million barrels per day (100 million tonnes a year). A fleet of 200,000 dwt tankers would be constructed to haul the oil across the Gulf of Alaska to Pacific Coast ports. Supplementary systems under consideration include crude oil and natural gas pipelines through Yukon Territory to Edmonton and then through existing systems to the mid-west. Natural gas is being moved from Kenai Peninsula and Cook Inlet fields to local Alaskan markets by pipeline, and crude oil and petroleum products to marine terminals for export, also by pipeline. Liquefied natural gas is being transported to Tokyo in two special refrigerator tankers, each with a capacity of 1.5 billion cu ft (42.5 million cu m).

The experience of the Second World War and the post-war period has demonstrated that air transport has shifted Alaska's strategic location. Nine international carriers operate North Pole routes between Japan and Europe through Anchorage. Alaska is no longer a remote land. The development experiences of the past decade within Alaska have further demonstrated that innovations in all forms of transport technology have eliminated or minimized what formerly were formidable barriers to access and movement. Alaska's natural resources will be developed or not in accordance with the state of our knowledge of their extent and nature, and of domestic and foreign market conditions. Geographically there is no longer (if there ever was) a development frontier in Alaska. Development could break out almost anywhere with little reference to other development or to surface transport systems.

National and international significance

Geographical location was the primary reason for the United States' purchase of Alaska in 1867. The events leading up to the Second World War and its aftermath were to demonstrate Alaska's high national importance. Situated on the Great Circle route between the continental land masses of the east and west, and at the crossroads of the North Polar air routes between Europe and the Far East, Alaska has become the United States' northern territorial foothold in Stefansson's 'Mediterranean' of the twentieth century. With the Canadian Arctic it has become the keystone of the North American air defence system.

The 1942 invasion of the Aleutian Islands by the Japanese underlined the importance of Alaska to national defence, and by the 1950s and 1960s Alaska had become primarily an 'exporter' of military defence, from the national point of view. Between 1951 and 1954, spending by the Department of Defense in Alaska averaged $412.9 million annually. The late 1950s saw the end of the build-up of the defence establishment to a plateau of maintenance and periodic renewal. Spending declined from a peak of $512.9 million in 1953 to annual amounts fluctuating between $264.6 million and $352.0 million during the decade of the 1960s. The number of military personnel stationed in Alaska, stabilized for the decade at about 33,000 service personnel and 6,500–6,800 civilian employees of the department, dropped to 25,300 military personnel and 5,400 civilian employees in 1975.

Although Alaska's strategic location is of continued importance to hemisphere defence and international travel and transport, national attention has steadily shifted to its natural resources. Gold and furs are no longer significant, but fish continues to give Alaska top ranking among the nation's fisheries, and there has been rapid growth in outputs of forest products and petroleum. Total natural-resource production rose from a gross value of $130.6 million in 1950 to over $600 million annually by the mid-1970s. These values are projected to exceed $2 billion when full North Slope oil production is achieved. A more comprehensive asessment of the net economic contribution of Alaska to the nation is given in calculations of Alaska Gross State Product (GSP) for the period 1961 to 1972 (Table

Table 22. Alaska Gross State Product, 1961, 1972, 1974 and projection to 1990

	Gross product in current (unadjusted) dollars			Gross product in constant dollars (1958 US PRICES = 100)			
	1961	1972	Average annual rate of change (%)	1961	1972	1974	1990[a]
	(millions of dollars)			(millions of dollars)			
All industries, total GSP	863.3	2,416.3	9.8	683.6	1,253.0	1,556.2	5,728.3
Government	304.0	791.8	9.1	256.2	298.8	309.5	467.3
Federal	248.5	490.6	6.4	225.1	212.5	217.8	217.8
State and local	55.5	301.2	16.6	31.1	86.3	91.7	249.5
Commodity-producing industries	229.6	688.2	10.5	171.5	393.5	562.3	3,321.7
Agriculture, forestry, and fisheries	27.5	50.0	5.5	30.9	20.3	20.0	37.2
Oil and gas	28.6	215.9	20.2	29.3	212.3	306.5	2,875.1
Other mining	8.5	8.1	0	7.6	6.8	7.0	7.5
Contract construction	81.7	271.2	11.5	33.6	61.4	98.6	158.2
Manufacturing	83.3	143.0	5.0	70.2	92.7	130.2	243.7
Support industries	329.7	936.3	10.0	255.8	560.7	694.3	1,939.4
Transportation, communications, public utilities	125.1	274.7	7.4	115.2	225.1	266.9	646.6
Trade, finance, insurance, real estate, services	204.6	661.6	11.3	140.6	335.6	427.4	1,292.8

[a] Based on intermediate petroleum development assumptions and oil at $45/barrel.
(Sources: University of Alaska, Institute of Social, Economic and Government Research, Alaska Review of Business and Economic Conditions, 'Estimated Gross State Product for Alaska', April 1974, Vol. XI, No. 1; and ibid., 'Alaska's growth to 1990, economic modelling and policy analysis for a rapidly changing northern frontier region', Jan. 1976, Vol. XIII, No. 1.)

22). This is a net value as it embraces only the value of goods and services produced within the boundaries of the State.

Before the Second World War, Alaska's foreign trade was negligible, consisting of goods moving through Alaska to and from Canada's northern territories. It remained insignificant until the 1960s, when exports directly to foreign countries increased in value from $6.5 million in 1959 to $129.1 million in 1970 and $264.9 million in 1975, most of this in shipments of forest products to Japan. Imports of pipe and industrial supplies and equipment from Japan also rose in keeping with resource development. Foreign imports expanded from $3.1 million in 1959 to $351.7 million in 1975. Interest is being shown in the entire range of Alaska's natural resources, including the purchase of fish and fish products from Alaska fishermen and shore plants, and there is significant international investment in many phases of Alaska's development. The growth of Alaska's foreign trade accompanying this shift in investment points towards growing international orientation.

Table 23. Alaska foreign trade 1931-75

Annual average	Value of imports[a]		Value of exports[a]	
	Total	Japan	Total	Japan
	(thousands of current unadjusted dollars)			
1931-40	335	c	378	c
1948-53[b]	1,417	c	2,830	c
1954-59	2,910	c	7,315	c
1960-64	7,050	800[d]	27,440	22,300[d]
1965-69	21,189	5,634	55,116	43,583
1970	78,372[e]	42,877	129,884	100,340
1971	41,274	19,860	126,116	105,771
1972	48,958	9,519	180,662	132,804
1973	39,878	6,805	205,050	178,805
1974	165,588[e]	37,545	260,571	192,582
1975	351,665[e]	152,984	264,942	191,591

[a] Imports do not include foreign products shipped to Alaska via the continental United States. Alaskan products shipped to foreign markets via continental United States ports are not included in exports, but goods transferred through Alaska from the continental United States to foreign markets are included.
[b] Does not include value of military and naval supplies and equipment shipped to Alaska.
[c] Data not available or amounts negligible.
[d] 1960 and 1961 estimated.
[e] Includes value of steel pipe and other supplies and equipment for construction of trans-Alaska pipeline from Japan and Canada (trans-shipments).
(Sources: US Department of Commerce, Bureau of the Census, Foreign Trade Report No. FM-563.)

The future of Alaska into the next century will be dominated by oil and natural gas. Projections in Table 22 by the University of Alaska's econometric models on the basis of intermediate assumptions as to the probable course and rate of development, show the Alaska Gross State Product in constant 1958 dollars increasing from $2 billion in 1974 to $6 billion by 1990. These models also project for the same period increases in monthly average employment from 165,000 workers to 356,700 workers and total resident population from 350,700 persons to 729,600 persons (over 1 million persons under the accelerated development assumptions). Most of the employment and population increase will take place within the two communities of Anchorage and Fairbanks because of the organization and operation of the oil and gas industry. These communities accounted for 59 per cent of the 1974 population (Anchorage alone for 44 per cent) and are projected in 1990 to account for 66 per cent of the total State population (Anchorage 54 per cent). This oil-and-gas 'Alaska' sometimes appears to be emerging as a separate entity, a northern suburb of Houston and other points in Texas, rather than as a part of the Alaska that was evolving prior to the international petroleum industry's arrival within the geographic boundaries of the State.

Table 24. Alaska State government revenue, 1961, 1972, 1974, and projection to 1990

Year	Total state government revenue	Total petroleum sector revenue[a]
	(millions of dollars)	
1961	45.7	4.2
1972	365.1	47.0
1974	428.7	48.0
1990[b]	4,840.3	2,339.0

[a] Direct oil and gas revenues only from production tax, rentals, royalties, and lease bonuses. Does not include corporate income tax.
[b] Based on intermediate petroleum development assumptions and $5/barrel.
(Source: University of Alaska, Institute of Social, Economic and Government Research, Alaska Review of Business and Economic Conditions, 'Fiscal data for Alaska', June 1975, vol. xii, no. 2; ibid., 'Alaska's growth to 1990, economic modelling and policy analysis for a rapidly changing northern frontier region', Jan. 1976, vol. xiii, no. 1.)

These projections also indicate that the most important impact upon the resident Alaskan economy and society will not be directly by petroleum-related activities, but through the revenues the State of Alaska and local government will receive from oil and gas leases, bonuses, royalties, and taxes. Table 24 projects under the intermediate assumptions that total State government revenues will increase more than ten-fold between 1974 and 1990 and that specific oil and gas taxes (excluding corporate income taxes) will account for 48 per cent of this ammount. Ironically, one of the biggest problems Alaska appears to be facing is what to do with all of this anticipated wealth. The course of reducing or even eliminating all taxes on other sources of income and wealth has been generally rejected on the grounds that it will only serve to attract greater numbers of people, which would tend to deflate the effects of these revenues on a *per capita* basis and create undesirable social and environmental conditions.

In November 1976 the voters approved the alternative of creating a permanent fund into which at least 25 per cent of all mineral-related revenues received by the State are to be deposited. The capital is not to be spent, but the legislature may make appropriations out of interest earned. On the assumption that all interest earned is spent each year, the fund balance could build up to $6 billion between 1977 and 1990. At 7 per cent interest the annual income from the fund would be $420 million. As oil and gas development and revenue will flow beyond 1990 and the fund balance continue to increase, the present economy of Alaska should have a life after death in the next century. It is the intention of the people that the annual income from this fund be used in ways that also will foster the rehabilitation of renewable resources, thus adding further support for a future economy when the petroleum resources run out.

An evaluation of Alaska's national and international significance, to be complete, must go beyond the calculation of regional gross product or trade statistics. Its contemporary and future significance, as discussed in this chapter, is emerging in the political context created by the Alaska Statehood Act of 1959 and the Alaska Native Claims Settlement Act of 1971. With its stated objectives of social and political as well as economic development, the 1959 Act decreed that the attempt be made to build a new American society or polity in Alaska. The 1971 Act not only recognized the hereditary claims of Alaska's indigenous minorities and attempted to make restitution for their infringements, but also decreed that these people shall be equal partners in this larger human experiment. Not content with high-sounding words alone, it also provided land, money, and corporate organization through which the search for true equality might effectively be conducted.

In an age when the nation faces the multiple crisis arising from excessive concentrations of population beyond the carrying capacity of the physical and human environment, and the conflict and violence of minorities to assert and preserve their identities, these experiments being conducted in Alaska take on a greater than regional or local importance. Their success or failure may produce models and guidelines for use in larger and more complex situations, and the fact that the experiments are being carried out in the north assures that they will be visible to the rest of the nation and the world. The general isolation of Alaska and its separation by foreign territory from the rest of the United States, the basic simplicity and its economic and social systems, the relative sparseness of population, all combine to present the essence of a laboratory situation in which results can be identified and cause and effect more readily traced than in the more complex situations found farther south. And as every northerner knows, it is very difficult indeed to keep any secrets in this country, and something quite like the truth will out.

CHAPTER FIVE

Greenland: the transformation of a colony

Geographical background

Greenland, the largest island in the world, has an area of 2,175,000 sq km. From the vicinity of Kap Morris Jesup (83° 39′ N), the most northern land in the world, to Kap Farvel (Cape Farewell) in the south is 2,670 km. The island is widest towards the north where the distance from the most western point to the most eastern is 1,050 km.

Greenland and North America were once part of the same tectonic plate, which split more than 50 million years ago along the line that is now marked by Davis Strait. Geologically, therefore, the island resembles nearby northern Canada, and Ellesmere Island in the Canadian Arctic Archipelago lies only 26 km away. Much of Greenland consists of a Precambrian shield of granites and gneisses, with mountains up to 3,733 m in the east, but lower in the west; in the north there are Paleozoic sedimentary plateaus.

Most of Greenland is still in the grip of an ice age. A great inland sheet of ice up to 3,500 metres thick rises to a maximum height of 3,300 m and covers four-fifths of the island. If this ice sheet were to melt, the sea level throughout the world would rise by about 6.5 metres. At Melville Bugt in the west and near Dove Bugt in the east, the ice sheet reaches the coast but elsewhere it is separated from the sea by a band of ice-free and frequently mountainous land up to 200 km wide, through which glaciers have cut deep valleys and narrow fjords. Many of these glaciers still descend to the sea. Some move at rates of several metres a day, and the crash as they calve to form icebergs often breaks the silence of the coast.

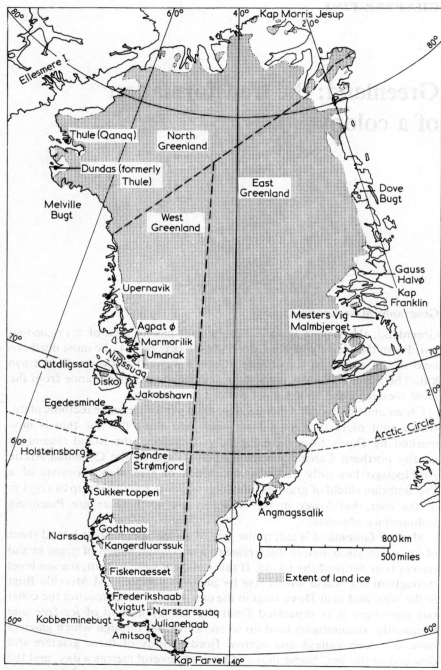

Map 16. Greenland

The ice-free land, amounting to nearly 350,000 sq km, is fairly evenly divided between the west, east, and north coasts. This ribbon of land, eroded by the ice as it cut its way through to the sea, is dramatic in its stark majesty. The mountains of the east and west coasts rise from long twisting fjords, combining and sometimes surpassing the beauty of the Swiss Alps and the Norwegian coast. On one side the ocean, dotted with dark islands and shining icebergs, stretches to the horizon. On the other side is the inland ice, white and featureless except near its edge where nunataks, the peaks of mountains penetrating through the ice, stand like rocky islands in a frozen sea.

The mean annual temperature on the Greenland ice sheet varies between $-20°$ C and $-30°$ C. In the winter the temperature can fall to $-70°$ C, and it rarely reaches $0°$ C even in the summer. Precipitation is least in the north and there is usually a wind blowing outwards and down from the highest parts of the ice sheet. Along the coasts the climate changes from polar continental in the north to maritime and quite mild for its latitude in the extreme south and southwest. Storms are common, especially in the winter on the southern coasts which are affected by depressions over the North Atlantic, and everywhere along the coasts the weather is very changeable. Katabatic winds from the ice sheet can sweep down the narrow fjords with little warning to cause strong local storms, though the weather may be calm only a few kilometres away.

The sea currents around Greenland have an important influence on the resources of the sea as well as on the weather and ice conditions. The East Greenland current flows south along the east coast, bringing fields of pack ice from the Arctic Ocean and making navigation very difficult even in the summer. Off southeast Greenland it is joined by the warmer Irminger current, a branch of the North Atlantic drift which bends back off southern Iceland to flow west. The difference in temperature between these two currents causes frequent fog. Mixing takes place as the currents pass Kap Farvel and turn north along the west coast of Greenland as the West Greenland current, warmer than the East Greenland current, and making the climate there much milder than that on the east coast. There is open water throughout the year on the west coast as far north as Holsteinsborg, except early in the year when heavy polar pack ice, known as the Storis, rounds Kap Farvel and spreads some distance north along the coast. The Storis can close harbours up to Frederikshaab throughout the spring. North of Holsteinsborg the winter is cold enough for sea ice to form in sheltered waters for several months, but the harbour of Holsteinsborg and the three ports to the south, Sukkertoppen, Godthaab, and Frederikshaab, are normally ice-free throughout the year.

Birch woods and mountain ash are found in sheltered valleys in the most southern part of the west coast of Greenland. Their growth is twisted and stunted and the tallest trees rarely exceed 6 or 7 metres. This stretch of coast, with its much less severe climate than the rest of Greenland, has extensive grassland and water meadows at the heads of the fjords, and it is here that the Norse colonists farmed and that sheep farming has been reintroduced today. Willow scrub, sometimes two to three metres in height, and mountain alder extend along the southwest coast up to about 71° N. Elsewhere in Greenland the vegetation is mainly Arctic in character, dominated by dwarf willow, heather, lichens, grasses, and a number of flowering plants which brighten the short summer.

Terrestrial mammals entered Greenland by the same route as the Eskimos, into the north of the island from the Canadian Arctic Archipelago. There are not more than nine wild species, and only two, fox and caribou, have had any significant direct economic importance. Muskox, which are protected, live in the north and northeast of Greenland where there are now several thousand, and a few animals have been introduced recently on the west coast. Caribou are found in West Greenland, and their numbers fluctuate greatly with short-term changes in the climate. When there are thaws followed by frost the ground becomes covered with ice which the caribou and muskox find difficult to penetrate to reach the lichen and grasses underneath. Early last century when conditions were favourable up to 40,000 caribou were killed in one year, but now in some periods they become very scarce. At present the annual kill is around 10,000 even though it is a good period. A herd of reindeer introduced from Scandinavia in 1952 has done well most years and has increased from 270 to several thousand, but few Greenlanders have taken up reindeer herding. Polar bear are most common in the north; about a hundred are killed each year. The Arctic hare occurs in north, northeast, and west Greenland though not on the southeast coast. Both blue and white Arctic fox are found throughout the country but trapping is no longer a significant source of income. Lemming and the ermine which prey on them live in the north and northeast of Greenland. The last pack of wolves killed was before the Second World War, and wolves have been very rarely seen since. Only a single wolverine has been reported in the last thirty years.

As summer approaches, flocks of waterfowl and other birds arrive from both Eurasia and America to breed. Some are killed for food, and large numbers of eggs are collected. Few remain in the winter except in the south. Of those that winter in Greenland the ptarmigan is the most important, and in good years they are very common. The eiderduck that breed in the north of the island spend the winter in southwest Greenland, which also receives large numbers of sea birds from northern Canada, Svalbard, and other northern islands. Many are shot by Greenlanders for food.

There are several hundreds of species of insects and of lower forms of life. Of these, midges, mosquitoes, and blackflies (*Simuliidae*) make their presence felt in the summer.

The seas of Greenland are more bountiful than the land, and man has been able to survive on the island mainly by hunting at sea. The currents in Greenland waters bring nutrients to the surface resulting in a rich growth of algae, the beginning of a food chain that leads to fishes and marine mammals.

Six species of seal are found off Greenland and all are hunted by the Greenlanders for their meat, oil, and skins. The commonest and smallest is the ringed seal which occurs on all the coasts. The much larger bearded seal is valued for its tough skin used for boots and skin line, and the scarce harbour seal makes the finest clothing. Hooded and harp seal are migratory and their numbers have been affected by hunting pressure in their breeding grounds outside Greenland. The largest of the seals, the walrus, appears to be very sensitive to disturbance and has been retreating to the north, but it is still important in the Thule area. The large Greenland whales were almost exterminated by commercial whaling in the eighteenth and nineteenth centuries, but may be recovering slowly. The smaller white whale, narwhal, and porpoise, however, remain numerous and are frequently hunted, as are minke whales.

Fish became increasingly abundant along the west coast early this century and fishing is now much the largest commercial activity. Several species have been exploited. For domestic use the Arctic char, found in rivers throughout the country, has always been valuable. The first commercial fishery for export was of shark, followed by halibut, and then cod. Since the Second World War shrimp, salmon, and especially cod have been the most important species.

In Greenland, where so much of life is marginal, climatic changes, however small, can have vital effects, greater than in many other parts of the world. The period when the Norse raised sheep in the valleys of southwest Greenland was followed by a colder spell. The Eskimos, living on sea mammals, replaced the Norse. Fish are very sensitive to changes in water temperatures, and the slightly colder climate of recent years has been followed by a sharp decline in the number of cod.

History

Man appears to have entered Greenland from North America by 2500 BC, and these first immigrants were followed by other groups from time to time. Several prehistoric cultures have been identified at sites along the coast, though much remains to be found out concerning their relationships, the areas they occupied, and when and why they succeeded or failed. The latest significant immigration was part of the vigorous expansion of the

Thule culture, which originated in Alaska towards the end of the first millennium and spread rapidly through northern Canada, across Smith Sound, and along the coasts of Greenland.

Nearly a thousand years have passed since the European discovery of Greenland, but after 1782 it was sheltered from the effects of European civilization until the outbreak of the Second World War. This was partly because of its inaccessibility and partly the result of deliberate policy on the part of the Danish government.

In 982 Erik the Red, outlawed from Iceland, sailed west to seek the islands rumoured to lie there. He found the coast of a new land which he called Greenland. Three years spent in exploration convinced him that it was a country that could support a colony. Back in Iceland he quickly gathered together a group of followers. In the late summer of 986 they reached Greenland, settling in the fjords of the west coast, which was then uninhabited though they found evidence that an earlier people had lived there. The Norse colony grew undisturbed for more than two centuries and the population probably reached about 3,000, but in the thirteenth century the settlers began to meet bands of Thule-culture Eskimos, or Skraelings as they were called, coming down the coast from the north, and there are references in both Icelandic annals and Eskimo tradition to hostilities between the two peoples. Meanwhile, the climate in Greenland was deteriorating, and throughout the fourteenth century conditions grew more difficult for the colony. Contact with Europe, devastated by the Black Death, became rarer and was finally broken in the fifteenth century. When it was restored by Frobisher in 1578, the Eskimos were in sole possession of the land. The fate of the Norse colonists remains a mystery. They may have fallen victim to raids by pirates; they may have been killed or absorbed by the Eskimos; they may have died from starvation or disease; some may have returned to Europe.

Frobisher was followed by other explorers and by Basque, British, Norwegian, and Dutch whalers. Many of the whalers, especially the Dutch, used to trade with the Eskimos along the coast each summer, exchanging mainly cloth, articles of metal, and, later, guns for blubber, whalebone, and skins. The native pattern of life must have changed in pace with a growing dependence on imported goods, including alcohol and tobacco. It was not until 1721 that a permanent European settlement was attempted. In that year Hans Egede, a missionary from Norway (which formed a single realm with Denmark), supported by some Bergen merchants and with the official blessing of King Frederik IV, established a station near where Godthaab now stands. Egede hoped to find some descendants of the Norse settlers and bring them into the Lutheran church, and to convert the Eskimos to Christianity.

The next fifty years saw the foundations laid for Danish sovereignty over

Greenland. Though disappointed at finding no descendants of the Norse, Egede pursued his mission of converting the Eskimos with great energy and zeal. He suppressed all Eskimo ancestral beliefs, looking on the Eskimo shamans as enemies, and succeeded in destroying their influence. An authoritarian hierarchy of traders and missionaries replaced native leadership and undermined the whole social structure of the natives, who also suffered the ravages of smallpox and other introduced diseases to which they had no immunity. Danish merchants, encouraged by the government, attempted to exploit the resources, but with little commercial success. Greenland was not as rich as they had hoped, and much of the trade remained in the hands of the whalers.

The Danish government, therefore, decided to take direct charge of both trade and administration. The Royal Greenland Trading Company (often abbreviated to KGH, standing for Den Kongelige Grønlandske Handel) was set up with a monopoly of all trade, and in 1782 Trade Instructions were issued closing the country to foreigners, giving the officials of the Royal Greenland Trading Company great authority, but placing restrictions on their actions. The guiding principle was that the welfare of the Eskimos should receive the highest possible consideration, even overriding where necessary the interests of trade. For administrative purposes the country was divided into South Greenland and North Greenland, with the boundary between the two at about 67° 30' N, and with each under an Inspector who reported to the Managing Director in Copenhagen.

The monopoly of the Royal Greenland Trading Company lasted for over 175 years, despite recurrent proposals in the spirit of nineteenth-century economic liberalism to open the trade to private enterprise. It gave Greenland a completely artificial economy which Denmark could and did manipulate to serve the interests of the policies she adopted. The price of imported goods could be adjusted selectively to encourage the purchase of useful articles and to discourage luxuries. Some could be prohibited, as was alcohol. The income of the Greenlanders could also be regulated by changes in the amount paid for Greenland products. Prices were determined annually, remaining the same throughout the year and throughout the colony, and this too could be done selectively to control the export of specific commodities. The price of sea-mammal oil and other products on the world market did, of course, have some effect, as it determined in the long run the amount that Denmark was willing to put back into Greenland, but the island was protected from short-term and capricious changes in world prices.

The long-term object of Denmark was to help the Greenlanders to attain a cultural and economic standard that would make it possible for them gradually to establish communication with the rest of the world without being exposed to exploitation. This implied education and the improvement of physical and social conditions. In the meantime the traditional hunting culture was to be retained and hunting techniques improved. Western civilization was to be held at bay until the Greenlanders were ready for it. It was an altruistic policy; it was also highly paternalistic, though maternalistic would be a more apt description. Its inherent weakness, of which Denmark became increasingly aware, was that it isolated the Greenlanders from the world for which they were being prepared, and there was no timetable to guide and test progress. It is difficult to learn to swim without getting into the water, and it is easy to put off making the attempt.

The Danish policy was not unsuccessful, especially in comparison with the policies or lack of policies in other Arctic areas. The rate of development in the economic, social, and political spheres was neither similar nor consistent, but over the years considerable progress was made. The isolation lasted until the Second World War, and the Greenland that then emerged was very different from the Greenland that had seen the beginning of Danish colonial rule.

Social conditions during the colonial period

The social anarchy that resulted from Egede's destruction of native leadership had far-reaching effects and did much to neutralize the well-meaning efforts of the Danish administrators. The sudden substitution of rules of conduct based on a different culture, under different conditions, in a different land, for the norms that had evolved over the centuries in response to the special conditions in Greenland, could lead only to confusion and demoralization.

The traditional communal way of life began to give way to a more individualistic pattern and a stratification of society slowly took place. About 5 per cent of the Greenlanders found employment with the Royal Greenland Trading Company or the missions. They followed European practices to some extent and enjoyed economic security. The great majority remained as hunters, attempting from depleted resources to feed and clothe their dependants, and to provide skins and blubber to the trading stations so that they could buy necessities, coffee, and tobacco. The situation was made worse by quarrels between the Danish Lutheran Church and the Moravian missions, which further confused the Greenlanders. The Moravians, who had come to Greenland in 1733, followed a policy of concentrating their followers around their missions, where the local resources could not support them and they became impoverished. By the middle of the nineteenth century distress was widespread throughout the colony. The people were ill-nourished, poorly clad, and living in insanitary conditions in sod and stone huts where the mortality rate was high, leading to a decline in the population.

In the second half of the century a number of far-reaching reforms were introduced, largely on the initiative of Henrik Rink, a Danish administrator, who recognized that the root cause of the malaise lay in the disruption of the traditional social system. Better housing was provided, a tax was levied on all products purchased by the Royal Greenland Trading Company to provide relief in emergency, and a monthly newspaper in Eskimo was initiated. A very important innovation was the establishment of Boards of Guardians which included native representatives elected by the Greenlanders. The Boards of Guardians administered relief and also had some judicial authority including the power to impose fines and banish people from certain districts. In this way a measure of responsibility for their own affairs was returned to the Greenlanders. Gradually both the morale and the standard of living of the Greenlanders rose and their numbers began to increase.

The improvement in conditions continued and accelerated during the twentieth century. In 1908 the Boards of Guardians were replaced by municipal councils which were composed at first entirely of Greenlanders, and were given more authority. These councils elected representatives to Provincial Councils, established for North Greenland and South Greenland, which advised the Inspector, or Governor as he was later called, of each of the two provinces. The Governors were responsible to the Greenland Department for the administration of their provinces. The Provincial Councils never met jointly until 1940, and co-ordination of policies in the two provinces was the responsibility of the Director in Copenhagen. Division of the country in this way ensured that control rested firmly in Copenhagen. Better education, which remained under the ecclesiastical authorities, and improved health services, played an important part in gradually increasing the general welfare.

During this colonial period Denmark extended her administration over all Greenland. She assumed responsibility for the Eskimos in the Angmagssalik area of East Greenland in 1894, ten years after their discovery, but it was not until 1921 that Danish jurisdiction was extended to cover the far northern Eskimos in the Thule area. In 1933 a long controversy with Norway over the sovereignty of northeast Greenland was finally settled in Denmark's favour by the International Court of Justice at The Hague, and Danish sovereignty over the whole island was secure.

The economy during the colonial period

At first trade in the young colony prospered, but the depression in Europe brought on by the Napoleonic Wars affected Greenland, which was largely cut off from Denmark, and conditions did not begin to recover until about 1825. A buoyant market for oil was a favourable factor until the last quarter of the century but from then on Greenland became an increasing expense. The population was growing, while the number of seals killed remained much the same in the south of Greenland and rose only slowly in the north.

Around 1910 the water of the West Greenland current began to get warmer and, though the rise in temperature was slight, it had far-reaching effects. Seals became scarcer first in the south and progressively farther north. By the beginning of the Second World War the annual take of seals was little more than half what it had been thirty years earlier. However, the change in the West Greenland current that caused the seals to retreat brought increasing numbers of fish, especially cod, to the west coast. Shark fishing had been making a useful contribution to the economy since the middle of the nineteenth century and the annual production of shark-liver oil had risen from 200 to about 700 tonnes. In 1911 a small cod fishery was begun. Fishing for cod expanded gradually until 1925 and then more quickly for a few years as the cod spread farther north along the coast.

In 1939, 10,000 tonnes of raw cod were taken and 2,300 tonnes of salt cod exported, a level that had then been maintained for ten years. In addition, the Greenland fisheries produced in the same year 100 tonnes of cod liver, 434 tonnes of shark liver, 6,700 shark skins, 24,400 wolf-fish skins, and 140 tonnes of halibut, with smaller amounts of other fish products. Fishing had become a well-established and profitable occupation, bringing a measure of prosperity to the island. As kayaks and row boats were mainly used, it was labour intensive and was restricted to close inshore. There was concern that the cod might depart as suddenly as they had come, and a resulting reluctance to make the heavy investments a modern fishing industry would require. This was not an unreasonable attitude as quantities of cod had appeared for a few years twice at least during the nineteenth century. In any case, before the Second World War Greenland was expected to pay for herself each year, and there was no way in which major long-term investments in her economy could be made.

Another response to the decline in seal hunting was the re-establishment of sheep farming. Sheep had been kept by the Norse colonists, but attempts to reintroduce them in Egede's time and towards the end of the nineteenth century had failed. In 1906, however, an experiment with some sheep imported from the Faeroe Islands had greater success. More sheep were brought into the country from Iceland, a sheep station was built at Julianehaab in 1915, and sheep farming became a new occupation. As would be expected in a marginal area where the weather was very changeable, there were heavy losses in some years but by 1939 the number of sheep was approaching 10,000. Small numbers of goats, cattle, horses, and poultry were also imported, but they were too few to play a significant role in the economy.

The Arctic fox was trapped everywhere in Greenland. They are not plentiful except in the north and attempts to farm fox and mink met with little success. Other renewable resources exploited were eiderdown, sea birds, ptarmigan, caribou, and polar bear. A few vegetables were grown in favourable areas in the southwest.

Non-renewable resources did little to help the economy with one important exception – cryolite. Greenland was fortunate in having at Ivigtut in the south of the island, much the larger of the only two deposits of cryolite known in the world, and this mineral became of commercial value at first in the enamel industry and later in the production of aluminium. Mining, by a private company, began in the middle of the nineteenth century and the annual output increased steadily throughout the years, reaching over 50,000 tonnes in 1939, in which year the Danish state acquired half the shares of the company. The mine did not affect the economic situation in Greenland directly as the Greenlanders were not involved in its operation, but the royalties and profits did much to reduce the expense of Greenland to Denmark. Other attempts at exploiting mineral resources were of lesser importance. Mining of coal began on the peninsula of Nugssuaq in 1905, and when this was exhausted a mine was opened in 1925 on the island of Disko, but the coal was of poor quality and only used domestically. Graphite, copper, and marble were also mined but on a small scale and did not prove profitable.

The modernization of Greenland

In April 1940 Germany invaded Denmark and the reins that had guided Greenland and had been held so long and so tightly were abruptly cut. They were quickly gathered up again by Eske Brun, then Governor of North Greenland. He took over the administration of all the island, consulting the two Provincial Councils on all important matters, and with Henrik Kauffman, the Danish Ambassador to the United States, arranged both for the purchase of supplies from North America and for markets for Greenland's products. An agreement was also negotiated with the United States for a number of defence installations, to help in the control of the North Atlantic and to provide an air route to Europe. Greenland was suddenly and unexpectedly brought into contact with the outside world, but the impact was not as damaging as it has proved in so many other countries. The great defence bases were carefully insulated from the rest of the island and did not require local labour or resources. Throughout the war the demand for both cryolite and cod was strong and these years proved a prosperous time for Greenland. A much wider variety of goods was introduced and there was money to pay for them. Politically it was a period of great significance, as the Greenlanders discovered that it was possible to exist without complete dependence on the mother country.

When the war ended Greenland was reunited with Denmark. Beneath the euphoria of renewing old contacts, there was some resentment among the more educated Greenlanders at the resumption of colonial status, and representations were made to abolish the monopoly of the Royal Greenland Trading Company, to be allowed more say in running their own affairs, to be treated equally with the Danes, and to have the administration based in Greenland rather than Copenhagen. These views were shared by many influential Danes who had been active in Greenland affairs. Denmark was not ready to move quite so fast but, following a visit by the Danish Prime Minister to Greenland in 1948, she appointed a Royal Commission composed of Danes and Greenlanders to advise on what changes should be made. On the recommendations of the Commission she began to speed development, abandoning the principle that Greenland should pay for herself.

During the next few years a major program of construction was undertaken and the Greenland Technical Organization was established to plan and carry out, in co-operation with private Danish firms, all technical assignments. Hospitals, schools, industrial plants, housing, and utilities were built, and many more students were sent to Denmark for higher education. The fishing industry was assisted with motor boats, docks, and harbours, extending its range beyond the inshore waters. A mine, financed by the Danish government and Danish, Swedish, and Canadian companies, was opened at Mesters Vig in East Greenland, where a deposit of lead-zinc ore had been found in 1948.

Meanwhile, a number of important acts were passed by the Danish Parliament. Greenland was divided into three districts: West Greenland, which combined the old North and South Greenlands, and included over 95 per cent of the population, North Greenland which covered the Thule area, and East Greenland. The Greenland Council was given a large measure of autonomy in West Greenland to provide a government very similar to that of a Danish county. Greenland ceased to be a closed country and was opened to influences from other lands. The trade monopoly was abolished by allowing Danish subjects, but not foreigners, to establish commercial enterprises in Greenland and to export the products of the country. The Royal Greenland Trading Company remained within the Ministry for Greenland but was separated from the administration, and education was secularized by being taken away from the control of the Church. A senior official of the Danish Department of Health was appointed Chief Medical Supervisor of the Greenland Health Service, responsible for the professional side of the medical and dental services. A special criminal code was adopted, the Danish Justice Department given responsibility for the police, and district courts were established for the administration of justice. In 1953 this evolutionary process reached its logical conclusion when the Danish Constitution was amended to include

Greenland as an integral part of the Kingdom of Denmark, electing two members to the Danish Parliament. Greenland was no longer a colony, and the Greenlanders had become fully fledged Danish citizens.

The following years saw tremendous efforts to modernize the country, a task made all the more difficult by the rapid growth of the native population, which had a natural annual rate of increase of over 50 per 1,000, among the highest in the world. Millions of kroner were spent annually in construction, education, health and social services, communications, and transport. Large numbers of Danish workmen were required, as the Greenlanders did not have the necessary skills. The fishing industry was seen as the most important continuing economic asset and one on which the modernization of the country could be based, and much of the activity was centred on the ice-free ports of the southwest coast. These were the main growth centres and all along the coast many of the small outposts were abandoned as the emphasis changed to fishing with motor boats working out of the larger settlements and selling their catch there. Many families moved from the seal-hunting area around Umanak and Upernavik to the fishing area farther south. This urbanization process was encouraged by the administration, not only on economic grounds but also because centralization greatly reduced the problems of improving health, education, and social services. In education the highest priority was given to teaching the Danish language, the Greenland school system was placed on an equal footing with that of Denmark, and the high schools were improved and expanded.

Efforts were intensified following the appointment of a new Commission in 1960, which in its *Greenland, 1960, Report* recommended greater rationalization of the fishing industry around the ice-free ports and a ten-year plan of very heavy investment not only in industry but also in education, health, municipal management, and other services in an effort to provide social conditions comparable to those in Denmark.

Despite the great progress that was made in raising living standards closer to Danish levels, it became evident during the 1960s that the objective of integrating the population was not being achieved. The pace and scale of development meant that much of the work was carried out by Danes, with the Greenlanders being left as lookers-on and recipients rather than as participants. Before the war there had been only some 200–300 Danes in Greenland, but with the modernization program their numbers increased until they formed a third of the labour force. The positions of responsibility and skill were filled almost exclusively by Danes. The situation was aggravated by a continuing problem regarding rates of pay. In order to attract Danish workers they had to be given greater remuneration than they would have received in Denmark. To pay Greenlanders at the same rates would have given Greenland pay scales well above those in the mother country. As a result there were two scales, one for Danes and the other, considerably lower, for Greenlanders, even when the work was the same. It was a dilemma that could not be resolved. The expanded education system, with half the native population under 14 years of age, and the greatly increased numbers of children attending school, required a great many more teachers from Denmark, who could not speak Greenlandic. The Greenlanders who had moved into the towns from the small settlements were faced with conditions that were strange to them. Private enterprise had been introduced to Greenland, but it was Danes rather than Greenlanders who had the training and experience to take advantage of these opportunities. As a result of such factors the Greenlanders began to resent the preponderance of Danish influence, and to conclude that the rate of progress had been too fast. It took some time for this feeling to be articulated, and it was in fact a reversal of the attitude they had adopted at the end of the Second World War when they had felt Greenland had been left too far behind and should catch up with the rest of the world with all possible speed.

In the last few years a Home Rule movement has taken shape and has quickly attracted wide popular support. Young Greenlanders who have received higher education in Denmark have been particularly active in promoting the concept. The national referendum on whether Denmark should join the European Economic Community showed how widely the interests of the two countries could differ. In Denmark as a whole the vote was decisively in favour but in Greenland it was 72 per cent against.

Denmark had had experience with nationalistic developments in Iceland and the Faeroe Islands, was quick to recognize the growing strength of the Home Rule movement, and accepted the inevitability of some form of devolution. Her response was to establish a Home Rule Commission, composed of equal numbers of Danish and Greenlandic politicians with a neutral chairman and an independent secretariat, to recommend how Home Rule should be brought about. The Commission was appointed in June 1975, and the Minister for Greenland announced subsequently that Home Rule would take effect from 1 May 1979. It is therefore in the context of Home Rule that the situation of Greenland should be considered (see also pp. 202–5 below).

Resources and their development
Fisheries
The plans for modernizing Greenland were built around the fishing industry. The appearance of large numbers of cod along the west coast early this century was seen first as a means of replacing the seal in the subsistence economy of the Greenlanders, and later as the economic

vehicle that could carry Greenland into the contemporary world. It was with this latter object that extensive efforts were made following the Second World War to rationalize and expand the industry. At first these measures met with considerable success.

The natural centres on which to concentrate the industry were the four ports of Frederikshaab, Godthaab, Sukkertoppen, and Holsteinsborg, as they were well spread along that part of the coast that was usually ice-free throughout the year, and the rich fishing banks lay only 50 to 100 km offshore. The building of fish-processing plants by both the Royal Greenland Trading Company and private enterprise, and the construction of quays and harbours meant that a greater tonnage of fish could be handled. The Greenlanders took advantage of the availability of loans and subsidies to obtain motor boats and quickly adapted their methods of fishing to make good use of them, extending their range of operations and greatly increasing productivity. They formed their own organization, the Union of Fishermen and Hunters of Greenland (KNAPP), to further their interests, and to prevent dominance of the industry by the Royal Greenland Trading Company. Production rose and in 1962 Greenlandic fishermen caught nearly 45,000 tonnes, a figure that has never been exceeded. Productivity in terms of weight of fish caught per fisherman also increased considerably. Up to this time the catch had appeared to be limited mainly by the capacity of the industry to secure and process the fish. Later in the 1960s, however, the catch declined despite continued technical improvements in the industry, and it became apparent that the cod were much less plentiful than they had been. There were probably two causes: one was a deterioration of the climate with lower sea temperatures off southwest Greenland, and the other overfishing. Numbers of large foreign trawlers were fishing in international waters on the banks off the coasts of Greenland, and were taking ten times as much fish as the Greenlanders could catch with their small inshore boats. Denmark therefore decided to extend the range of the Greenlandic industry by building trawlers which could fish the offshore banks like the foreign trawlers, and would also be able to go farther afield if the cod were to retreat from the Greenland waters. In this way Greenland, which in 1974 had seven trawlers, has managed to retain a substantial share in the catch, but the cod fishery, on which such high hopes had rested and so much money had been spent, has proved a very great disappointment.

Before and immediately after the Second World War, cod had constituted almost the entire catch, but in recent years other fishes have represented about half. There has been a marked decline both in the Greenlanders' take of cod, which has fallen to only half that of the record year of 1962, far below the capacity of the industry, and in the number of cod taken by the foreign trawler fleet on the Greenland banks. By 1973 the total Greenland catch of fish was less than 40,000 tonnes with the proportion of cod decreasing, and it remained at very much the same level in 1974 and 1975. On the other hand, the catch of shrimp has been steadily growing following the discovery of exceptionally large shrimp beds in 1948 in Disko Bugt and more recently off the west coast. Some 10,000 tonnes of shrimp are now produced each year and this can probably be increased, especially if the catch by countries outside the European Economic Community is reduced by the establishment of the 200-mile limit. There is however growing competition especially from Alaska, where shrimp beds have been found even richer than those in Greenland waters. The other fish that has made a very substantial contribution to the economy in recent years is salmon, but fishing for salmon is causing serious international problems. There is widespread concern that the species could be in danger. The catch is now under a quota and it would be unrealistic to consider salmon to be a resource that could be exploited to a greater extent. Wolf-fish, halibut, and redfish are the most important of the other fish caught by Greenlanders.

The Greenland fisheries are at present by far the largest factor in the economy of the country, employing about 3,000 fishermen, though some only occasionally, and up to around 1,500 workers at the fish-processing factories. In 1973, the value of fish sold to the Royal Greenland Trading Company and to private exporters was 60.2 million kroner, out of a total of 69.9 million kroner for all purchases of Greenlandic renewable resources. Cod accounted for 17.8 million kroner, salmon for 18.6 million kroner, shrimp for 14.9 million kroner, and all other fish for 9 million kroner.

Recent declines in both the catch and export prices combined with increased operating costs have caused serious financial difficulties for the private fish-processing plants. There are thirteen private companies, four of which are co-operatives. They handle a quarter of the catch, but as this is mainly the more expensive products – shrimp, salmon, halibut – it represents a third of the value. A temporary subsidy based on the weight of fish purchased was paid in 1975, and long-term subsidies and other financial assistance will be required if the companies are to be kept in operation as an alternative to state-operated plants. Some of them are important employers and some provide the only fish-purchasing agency in a community. In certain cases the companies may have to be taken over by the Royal Greenland Trading Company.

The Royal Greenland Trading Company has been seriously affected by the same factors and has had an increasing deficit on its production activities since 1965, when production balanced costs. By 1975 the annual deficit amounted to about 70 million kroner.

The future of the industry would appear to depend on two factors. One is the recent establishment of the exclusive economic zone extending to 200 miles offshore – the area where fishing is controlled by the littoral state. This would give Greenland exclusive fishing rights on the rich fishing banks off her southwest coast, where foreign countries have been catching hundreds of thousands of tonnes of fish a year. As a member of the European Economic Community these rights would be shared with other members. The other factor, therefore, is what arrangements Denmark can make within the European Economic Community to afford protection to the Greenland fisheries against competition from other members. In a Home Rule situation it is likely that Greenland will elect to leave the European Economic Community. She will still however have to share the fishing off Greenland with other countries as her catch alone is not as great as the fishery could sustain.

Hunting

Hunting, the traditional pursuit of the Greenlanders, remains the most important calling in North and East Greenland. In West Greenland the growth of the fishing industry has to a considerable extent been at the expense of hunting, as many people from the northern hunting areas centred on Upernavik and Umanak have moved south to the fishing ports. The number of sealskins purchased annually by the Royal Greenland Trading Company has, however, actually increased in the last forty years from 25,000 just before the war to a current figure of about 50,000, about one-third of which are from East Greenland and under 10 per cent from North Greenland. The increase in the number of sealskins is partly the result of higher prices and partly because the Greenlanders no longer require so many for their own purposes for clothing, tents, and covering boats. The seal is also a valuable source of meat. In contrast to seal hunting, fox trapping has declined over the years and the annual number of blue and white foxes purchased has decreased from about 5,000 before the war to below 2,000. In 1973 the value of animal skins produced in Greenland, the great majority of which were sold to the Royal Greenland Trading Company, was 7.7 million kroner, or 11 per cent of the total value of the purchases of Greenland products.

Hunting has had a long history in Greenland and it will probably have a long future. It survived despite the official emphasis on fishing and urbanization. Though there has been little movement yet back to the smaller settlements, the drift to the towns seems to have been arrested and the plans of the Greenland Council to resume expenditures at the outposts rather than to concentrate services in the main centres will make life there more attractive. The growing interest in their own culture and the nationalism that will be fostered by Home Rule, are likely to encourage seal hunting, and some modest expansion is probable over the next few years.

Farming

The only type of farming that has proved practicable in Greenland has been sheep farming in the extreme south of the west coast, and even here it is marginal owing to the climate. Winter feed is also a major problem owing to the restricted area of land that can be cultivated. In most years conditions are satisfactory and good progress is made but it can all be undone by a particularly severe winter causing great losses.

During the war years the number of sheep increased from 10,000 to 17,000, and by 1966 it had reached 45,000, a figure that has not been exceeded. Unless the climate deteriorates, sheep farming will continue to make a substantial contribution to the economy. It is the main support of some 200 families and is a useful addition to the Greenland diet. In 1973 the value of sheep products sold to the Royal Greenland Trading Company was 1,936,000 kroner. Climatic constraints and the limited area suitable for sheep will, however, prevent any major expansion, and 60,000 sheep is probably the maximum that could be carried.

Tourism

Greenland is attractive to tourists on many counts. The scenery is unsurpassed, the Greenlanders are colourful, friendly, and interesting, the flora and fauna are distinctive, and the peace and quiet offers a refreshing contrast to the pace of life in the developed countries. Tourism has often been suggested as a business that could be greatly expanded to broaden the economy of the country.

There are, however, a number of serious problems associated with the development of tourism in Greenland. Hotel accommodation, at present very limited and of modest standards, would have to be greatly expanded and improved. As there are no highways, internal transportation, for both excursions and travelling from one stop on the itinerary to the next, would have to be by water, which is slow, or by air, which is subject to frequent unpredictable delays and is expensive to the tourist and also to the government, which has to subsidize the service. As the season is short, the facilities established for tourists would be fully used for only a few months of the year and employment of Greenlanders would be mostly in the summer months, when unemployment is lowest. There is little for tourists to do in the evenings, one settlement appears to be much like another, there are few museums or historic sites, and a major influx of tourists could destroy much of the character of the country. Cruise ships add little to the local economy and carry too many people in relation to the size of communities in Greenland. Even at present, the tourists complain of the lack of handicrafts and other local products they can buy. The greatest deterrent is probably the high cost of travel to Greenland which puts a holiday there out of the reach of many who might like to visit the island.

The total number of tourists, including campers and one-day visitors, has not changed much in the past few years. In 1972 there were 6,500, in 1974 the number had fallen to 6,100, and in 1975 it recovered to 7,000. The number of campers rose from 300 in 1972 to 1,400 in 1975. It seems unlikely that there will be any sudden surge in tourism, but there are many to whom Greenland will always have a special appeal. They include anglers, mountain climbers, and those who enjoy nature. The number of visitors can be expected to show a gradual increase, which can be handled without unduly disturbing the character of the country. The rate of growth will depend to a considerable extent on whether a regular direct air service is established between Greenland and North America, where many of the potential visitors live.

Mineral development

With the exception of cryolite from Ivigtut and lead and zinc from Mesters Vig, exploitation of mineral resources in Greenland has been on a very small scale. The post-war modernization program has been based exclusively on the fisheries, partly because fish were a known major resource and partly because fishing was believed to be an occupation that the Greenlanders could follow with less adjustment and less specialized training than mining, which is becoming increasingly dependent on large-scale operations involving automation and sophisticated equipment. When the coal mine on Disko was closed in 1972, there were no mines in operation in Greenland for the first time since the cryolite mine opened in the middle of the nineteenth century.

The Danish government had, however, already taken measures to encourage the search for minerals in Greenland. There were several reasons for this decision. Relying on the fisheries alone was too risky as it made the whole economy vulnerable to changes in a single market. Even more serious was the danger that the fish stocks would decline. This had always been a source of concern, and the decreasing catches of cod after 1962 showed that these fears were well founded. It was also a time when interest in non-renewable resources was growing rapidly in other polar countries.

As Denmark herself had little experience in mining, the Ministry for Greenland made a careful study of how the industry operated and was regulated in other northern countries. A bill was then introduced in the Danish Parliament covering fundamental principles to govern mining in Greenland and establishing a general framework around which regulations could be fashioned.

In May 1965 the 'Act on Mineral Resources in Greenland' was passed by the Danish Parliament. It reaffirmed that all mineral resources were the property of the state, but allowed the inhabitants of Greenland free use of coal, peat, soapstone, gravel, etc., in accordance with prior practice. It divided the development of mineral resources into three stages: prospecting – to locate the presence of mineral resources, exploration – to determine the extent of a resource in order to see if it could be exploited, and exploitation – to mine and move the resource to market. The Act laid down the conditions under which the Ministry for Greenland would grant prospectors' licences and concessions for mineral exploration and exploitation. The prospectors' licences, which are for a maximum period of five years, do not confer exclusive rights over a specific area or any right to a subsequent concession for exploration or exploitation. The exploration concessions have a maximum life of eight years in West Greenland and twelve years in North and East Greenland, provide exclusive rights for a specific area, and require exploration to begin as soon as possible. They also carry the understanding that the holder of the concession will be given priority if, as a result of his exploration, he subsequently applies for a concession to cover exploitation. The exploitation concession provides exclusive rights over a specific area for a maximum period of fifty years and can be issued only to companies registered in Denmark. The exploitation concession has to cover in detail such matters as taxation, safeguards against pollution and other environmental damage, employment of Greenlanders and Danes, and Greenlandic and Danish participation in contracts for supplies and services. It also specifies the ways in which the state will share in the profits of the operation after the return of invested capital with appropriate interest. This may be by royalties, by fees, or by interest on shares, and will probably amount to between 35 and 55 per cent of the net proceeds. An exploration concession has to be approved by a special Parliamentary Committee and the opinion of the Greenland Provincial Council has to be obtained. The operating plans for exploitation prepared by the companies require the approval of the Ministry for Greenland.

The legislation succeeded in attracting a number of companies, mainly foreign, who began to investigate the mineral possibilities. Within four years prospectors' licences had been issued to five companies, nine exploration concessions had been granted, thirteen applications were under consideration at the Ministry for Greenland, and a number of applications had been refused.

In 1971 the first exploitation concession was granted. It was to the Greenex Company for the exploitation of minerals, excluding hydrocarbons, cryolite, and radio-active substances, in an area near Umanak, where the 'Black Angel' lead-zinc deposit had been found. In practice companies have been reluctant to undertake the expense involved in exploration without knowing the conditions under which exploitation would be allowed. There is no reason under the act why the exploration and exploitation concessions cannot be combined and this has been done on occasion.

Interest in the possibility of *petroleum* in Greenland first developed soon after the Second World War and was centred in the Disko Bugt area. The results of preliminary exploration were not promising enough to attract support and it was discontinued. At the time the legislation was being drafted the main non-renewable resource potential was believed to lie in metallic ores on the mainland, and little or no consideration was given to either petroleum exploration or the continental shelf. The oil crisis and its effect on world petroleum prices, the discovery of the very large oilfield at Prudhoe Bay on the Arctic coast of Alaska, and the unexpected successes of offshore drilling in the North Sea, refocused interest in the possibility of finding oil in Greenland. The continental shelf is extensive off both the east and west coasts of the island and the geology is believed to be favourable.

Petroleum exploration and offshore operations both entail very heavy costs in the exploration stage, but the legislation was sufficiently flexible and in such general terms that it could handle the new conditions with only minor amendment. A model concession to explore for and export petroleum was drawn up by the Ministry for Greenland and agreed to by both the Special Parliamentary Committee and the Greenland Provincial Council.

The model concession gives sole and exclusive rights to explore for and exploit petroleum in a defined area and to sell and transport the petroleum obtained. Operations have to be as expeditious as possible and designed to obtain full exploitation. Drilling programs must be approved by the Ministry for Greenland and the results reported to the Ministry, which reserves the right to inspect the operations. Compliance with Danish laws is required and pollution safeguards are specified. Contracts and the provision of supplies and services are to be assigned to Greenlandic and Danish enterprises if competitive, and Greenlandic and Danish personnel must be used wherever possible, and trained for operations and management.

The concession covers exploration for a period of ten years which may be extended to sixteen. A minimum rate of expenditure is required each year. At the end of six years a third of the concession area must be surrendered and another third at the end of ten years. If an exploitable field is discovered, exploitation is for a further period of thirty years which may be extended.

During the exploration period there is an annual charge per square kilometre of the concession area, and this increases if commercially exploitable deposits of petroleum are discovered. The royalty payment, which Denmark can take in the form of petroleum, is 12½ per cent of the value of all petroleum produced or wasted by negligence, and there is also an annual charge during production equal to 55 per cent of the profits, from which Danish company taxes can be deducted. The Ministry for Greenland reserves the right to share in the development and exploration of the field up to 50 per cent. It may also require the petroleum produced to be delivered for consumption in Denmark.

The Danish government decided that the first concessions to be granted would be in the sector of the continental shelf off the west coast of Greenland south of 70° N and the area was divided into blocks of approximately 400 sq km. Only some of the blocks in this area would be allotted initially.

Following the publication of the draft concession, a number of oil companies made applications and in April 1975 concessions were signed with six consortia, each made up of three or more oil companies. A total of forty-six blocks were allotted in thirteen separate concessions, some of the consortia being awarded more than one. One consortium began exploration drilling in 1976.

The award of these concessions, though agreed to by the Greenland Provincial Council, was not welcomed by all Greenlanders. Some were apprehensive about the effect that oil exploration might have on the fisheries and some considered that no action should have been taken until the advent of Home Rule. There was also concern about the possible impact of offshore drilling on the fisheries and on the ports of the west coast where much of the supply and servicing of exploration activities would necessarily be based and where considerable numbers of well-paid foreign workers might be expected.

Petroleum could prove to be by far the most valuable of Greenland's non-renewable resources. If exploration succeeded in finding oil in commercial quantities it could revolutionize the economy of the island.

All the *cryolite* at Ivigtut has now been mined, but there is said to be sufficient stock-piled to supply normal shipments until about 1985. The total output from this unique deposit, which was mined for over a century, amounted to about 3.5 million tonnes.

Uranium ore was discovered in the early 1950s in the Ilimaussaq intrusion on the Kvanefjeld plateau, 16 km from Narssaq, but the grade was too low for it to attract much attention at the time. Some drilling was however carried out during the next few years. The recent concern about energy resources, the limited amount of uranium that has been found in relation to the estimated future world consumption, and the increased price of uranium have revived interest in the deposit. As a member of the European Atomic Authority, Denmark is also committed to carrying out adequate exploration for uranium.

About 32 million tonnes of ore, containing an average of around 300 grammes of uranium per tonne are known and further exploration will certainly increase this figure. Efficient extraction would require a large-scale operation treating some 500 tonnes of ore an hour. A pilot plant handling 100 kg of ore an hour is being built to test the feasibility of production.

Indications of uranium have been found at two other locations in Greenland, one in the Søndre Strømfjord area and the other at Kap Franklin on Gauss Halvø in East Greenland.

The *lead and zinc* deposit at Mesters Vig in East Greenland has been worked out. During its short life of eight years it produced 58,000 tonnes of lead concentrates and 75,000 tonnes of zinc concentrates, the proceeds from which just about balanced the cost of production.

The only mine in operation in Greenland in 1976 was the 'Black Angel' mine at Marmorilik east of Umanak, where lead-zinc ore is recovered from a mine in the side of a high cliff, accessible only by cable car. The mine is the property of Greenex, a wholly owned subsidiary of Vestgron, which is controlled by Cominco, itself a subsidiary of Canadian Pacific. The 'Black Angel' mine went into production in the fall of 1973, and extracts about 2,000 tonnes of ore a day. Known ore reserves at the end of 1975 were 4 million tonnes containing 4.9 per cent lead and 14.1 per cent zinc. The company has also investigated a zinc ore deposit on Agpat Ø, 30 km from Marmorilik. Net earnings were 76 million kroner in 1974 and 91 million kroner in 1975, and the long-term debt had been reduced to 194 million kroner by the end of that year. When capital expenditures have been recovered, a royalty of 45 per cent of the subsequent profits will be paid to the Ministry for Greenland. The government benefits also from taxation, including income tax on the pay of the employees at the mine. Owing to an oversight in drafting the Greenlandic tax law, the municipality of Umanak, which has a normal annual budget of about 500,000 kroner, received an additional and unexpected 10 million kroner in 1975 from this source.

The mine employs about 325 men, most of whom are Danes but between 30 and 40 are Greenlanders. The labour turnover has been high especially among the Greenlanders, who do not object to the work, but resent the fact that they are paid much less than the Danish workers receive for similar work.

Chromium ore deposits have been reported over a large area in Fiskenæsset but the ore occurs in thin layers. The richest deposit has about 2.5 million tonnes containing some 350,000 tonnes chromium oxide. Other known deposits in the area have over two million additional tonnes of ore.

Copper was one of the first minerals to be exploited in Greenland. A small mine at Julianehaab yielded some 15 tonnes of copper in the middle of the nineteenth century, and early in the present century a few tonnes of ore were extracted from Josva and Lillian mines in Kobberminebugt (Coppermine Bay). However, although copper mineralization has been reported at several places on the island, it has not been found in sufficient amounts to justify exploitation under modern conditions.

An important deposit of *molybdenum* ore has been found in East Greenland at Malmbjerget south of Mesters Vig. The deposit has been estimated to contain nearly 120 million tonnes of ore with an average of 0.25 per cent molybdenum sulphide. The very expensive transport system that would be required from the mine to the coast, and the high cost of the energy needed to extract and concentrate the ore are the main factors that have prevented exploitation.

An *iron* ore deposit estimated at 2 billion tonnes of ore with an average iron content of 38 per cent has been located near Isukasia in Godthaab Fjord. An ore of this low grade could be profitable only if worked on a very large scale. This would require massive capital expenditure and would result in the establishment of a new community of at least 4,000 people. The economic feasibility is now being examined.

The *coalmine* at Qutdligssat on Disko was closed in 1972 as it had never proved profitable and coal was losing ground to oil. The mine was on a stormy stretch of coast that had no harbour and the coal had to be lightered out to ships. Between 1960 and 1968 annual production had ranged between 20,000 tonnes and 39,000 tonnes and the total expenses had been 82.9 million kroner against an income of 58 million kroner. The mine had, however, provided employment for a population of about 1,000. The settlement has now been abandoned. There are extensive coalfields on the peninsula of Nugssuaq in Disko Bugt, but they have not been fully explored. The coal, which is of poorer quality than the coal imported into Greenland, occurs in thin seams which proved to be discontinuous as the area has been faulted. The feasibility of exploiting this coal in the future will depend largely on the price of oil and nuclear power.

Graphite occurs at a number of places and there has been some extraction in the past, but it did not prove profitable. The Amitsoq mine in South Greenland yielded some 6,000 tonnes of graphite before it closed in 1925.

Marble has been quarried at Marmorilik near Umanak and some was exported to Copenhagen. The operation has now been discontinued.

Zirconium ore running up to 4 per cent zirconium oxide has been found at Kangerdluarssuk south of Narssaq. The richer ore contained about 60,000 tonnes of zirconium oxide but there is much more ore of lower grade.

A number of other mineral occurrences have been reported in Greenland. They include tungsten, nickel, platinum, beryllium, thorium, niobium, silver, and a few small diamonds. Some of these discoveries are being explored further but there are as yet no plans for exploitation. Interest in Greenland's mineral potential continues however and the number of prospecting licences issued is increasing. This is sure to result in more discoveries.

Hydroelectric power

As there are no large rivers in Greenland conventional hydroelectric plants have not been possible. The Ministry for Greenland has recently, however, requested the Greenland Technical Organization to initiate an investigation into the feasibility of a proposal to develop a plant based on a system of high-level lakes at the edge of the ice cap at Narssarssuaq. This would serve the town of Narssarssuaq, and might be extended much later to Narssaq and Julianehaab, but a major industrial use of power would be required for it to be competitive with plants fuelled by coal or oil. If the uranium ore found at Kvanefjeld can be exploited economically, the mine would be an important consumer of power. There are similar lakes elsewhere in Greenland that could be developed, and some are being looked at in detail.

Population

When Egede arrived in Greenland in 1721 there were possibly about 8,000 Eskimos living there, but the first fairly accurate census of the west coast was not until 1805. In that year the population was said to be 6,046. This did not, of course, include the far northern Eskimos in the Thule area or those on the east coast, as neither group was then known, but they would not have added more than a few hundred. Throughout the colonial period the population increased, gradually at first with occasional setbacks, but more rapidly during the twentieth century, and by the outbreak of the Second World War it was approaching 20,000 (Table 25). The population was about three times as great as it had been a hundred and fifty years earlier and this increase had taken place during a period when Eskimos elsewhere in the world had been declining in number. It was no longer a race of pure Eskimo stock, as there had been considerable infusions of foreign blood, first from the nations that had taken part in whaling off the Greenland coast, but especially from the Danes sent out

Table 25. Population of Greenland, 1805–1975

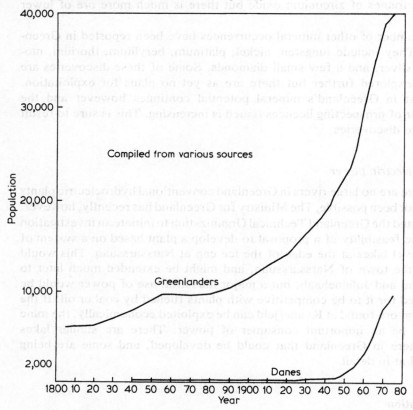

Notes: East Greenland included from 1901 (then under 500). Thule area included from 1921 (then 250).

to Greenland, many of whom had married and raised families in the country.

During and after the Second World War the population continued to grow at a rapidly increasing rate. In the twenty years following 1950 it very nearly doubled. There were two main reasons for this population explosion. One was the great improvement in health services, leading to reduced infant mortality and a longer expectation of life. The other was the massive immigration of Danes needed to provide improved services and to assist in carrying out the modernization program. Up to the end of the Second World War the number of Danes in Greenland had remained well below 500, but by 1973 it exceeded 9,000.

By the end of 1973 the total population had reached 49,468, the great majority living along the west coast. West Greenland had a population of

44,805, or 92.2 per cent, while the figures for East Greenland were 3,068 and for North Greenland 708. Within West Greenland the post-war years have seen great changes in the distribution of the population. When hunting was the main occupation the people were spread along the coast in small settlements. The growth of the fishing industry led to considerable migration from the hunting districts, particularly to the year-round open-water ports of the west coast, a movement that was encouraged in every way by the administration. There was a need for labour at the fish-processing plants and it was at the main fishing ports that the massive construction of the modernization program was concentrated and that new houses, health services, schools, and other facilities were being built. Table 26 shows the occupations followed by the population of Greenland on 31 December 1970. It illustrates the dominating position of the fishing industry in the economy, and of government and state enterprises as employers.

For many years the amount paid to fishermen for their catch was higher at the fishing ports than at the smaller places, many of which were abandoned as the people moved into the towns. By 1974, 34,894 people were living in towns in West Greenland and only 8,529 in the smaller settlements. Godthaab had a population of 8,258, Holsteinsborg 3,716, Egedesminde 3,379, Jakobshavn 3,200, Julianehaab 2,879, Sukkertoppen 2,889, and Frederikshaab 2,393.

The past few years have seen growing resistance in Greenland to urbanization, and people have questioned the wisdom of encouraging the movement towards the fishing towns at a time when the fisheries are becoming less productive. The smaller places had been neglected in the modernization program but efforts are now being made to improve health conditions there, with better housing and other amenities, and the drift to the towns will probably be reduced.

No less dramatic than the post-war population explosion has been the subsequent sudden fall in the Greenland birth-rate, which dropped from 49 to 19 per thousand between 1964 and 1972 with the introduction and encouragement of contraceptive measures. As a result the growth of the Greenland population will be much slower than had been expected, but it will still be considerable because of the increasing number of women now entering the child-bearing age.

The recent changes in the rate of growth have led the Greenland Council to forecast a total population in 1985 of 50,000 which is a reduction of 5,000 from the forecast made only three years earlier. They now expect the towns, especially the main fishing ports, to grow much less rapidly, and that the migration from the hunting districts and from the smaller places will be more than compensated for by the natural growth of the population there.

It is very difficult to forecast the size of the non-native population. The

Table 26. Population of Greenland by occupation, 31 December 1970

Category	Fishing hunting farming	Mining	Manu-facturing	Construc-tion	Utili-ties	Commerce	Transport	Admin; professional service industries	Old age and other pensions	Other	Total
Economically active											
Self-employed	2,393	1	77	297	—	100	115	85	—	—	3,068
Working within family	6	—	1	1	—	20	3	8	—	—	39
Fishermen receiving shared catch	1,016	—	—	—	—	—	—	—	—	—	1,016
Salaried employees, public servants, etc.	6	22	189	382	50	1,264	675	2,608	—	52	5,248
Royal Greenland Trading Company	4	—	98	—	—	837	150	31	—	—	1,120
Greenland Technical Organization	—	15	24	222	47	—	233	20	—	—	561
State government	—	—	12	—	—	20	117	2,158	—	—	2,307
Local government	—	—	7	2	1	—	—	160	—	—	170
Private sector	2	7	45	158	2	403	121	210	—	—	948
Unknown	—	—	3	—	—	4	54	29	—	52	142
Wage earners	72	131	2,104	2,073	191	682	1,069	2,253	—	294	8,869
Royal Greenland Trading Company	11	76	961	—	—	121	208	540	—	—	1,917
Greenland Technical Organization	—	—	216	365	145	—	214	71	—	—	1,011
State government	—	—	66	13	8	334	411	540	—	—	1,372
Local government	—	15	34	7	18	—	16	254	—	2	346
Private sector	57	39	681	1,670	17	227	148	783	—	46	3,668
Unknown	4	1	146	18	3	—	72	65	—	246	555
TOTAL	3,493	154	2,371	2,753	241	2,066	1,862	4,954	—	346	18,240
Men	3,479	146	1,634	2,650	239	1,278	1,714	1,541	—	200	12,881
Women	14	8	737	103	2	788	148	3,413	—	146	5,359
Others											
Housewives	1,308	63	393	676	89	453	500	501	524	17	4,524
Children not working	5,207	226	2,087	2,679	352	2,295	2,230	3,396	1,862	165	20,499
Children helping at home	87	5	18	14	5	25	11	31	58	—	254
Maidservants	18	1	11	25	3	24	26	98	40	1	247
Old-age pensioners, etc.	—	—	—	—	—	—	—	—	2,767	—	2,767
TOTAL	6,620	295	2,509	3,394	449	2,797	2,767	4,026	5,251	183	28,291
Men	2,639	106	1,050	1,364	173	1,175	1,122	1,714	2,110	82	11,535
Women	3,981	189	1,459	2,030	276	1,622	1,645	2,312	3,141	101	16,756
All population											
TOTAL	10,113	449	4,880	6,147	690	4,863	4,629	8,980	5,251	529	46,531
Men	6,118	252	2,684	4,014	412	2,453	2,836	3,255	2,110	282	24,416
Women	3,995	197	2,196	2,133	278	2,410	1,793	5,725	3,141	247	22,115

(Source: *Statistisk Årbog, 1976*, Copenhagen, Table 412.)

available statistics do not include military personnel. The United States base at Thule must at one time have had a strength of several thousand but it has been greatly reduced in recent years. In any event the Thule base, the Distant Early Warning Line stations, and other defence installations have remained isolated from the Greenlandic population and it is Danish contractors and their Danish employees that supply many of the services.

With increased education and training, Greenlanders will certainly be in a position to replace a number of Danes and the proportion of Danes in the population can be expected to fall, especially as the government construction program has passed its peak. However, any new mining developments will depend largely on imported labour, initially at least. Similarly offshore oil exploration and the shore services needed to support it will require men with highly specialized technical training who cannot be supplied from Greenlandic sources.

Home Rule in Greenland is likely to accentuate the replacement of Danes by Greenlanders. It will also probably discourage the emigration of Greenlanders to Denmark, which had appeared to be the only solution for the increasing population and which the government had been prepared to assist.

Social conditions and health

The educational system in Greenland before the Second World War had been designed for a population that was isolated from the rest of the world and depended on its own resources. It was ill-suited to respond to the outside world to which Greenland was introduced. Until 1950 education had remained in the hands of the Church. The great majority of the Greenlanders had become literate in their own language and, except in the Thule area and East Greenland, this had been the case for nearly a hundred years. A small number of Greenlanders had proceeded through high school to be trained as teachers or to occupy positions in the government or the Royal Greenland Trading Company. A handful had attended university. This was not the sort of system that could withstand the strains of modernization, or provide the educational needs of an increasingly technological age.

In 1950 control of all educational matters was transferred to a Central Board of Education which consisted of the Director of Education, who was responsible for the administration and operation of the schools, the Dean of Greenland, to provide continuity with the old regime, and the Governor of Greenland as Chairman. Local education districts were organized, school facilities were greatly improved, and evening classes were initiated, at first mainly for traditional subjects and later for voca-

Table 27. Number of pupils in schools in Greenland, 1951–69

	1951	1954	1959	1964	1969
Towns	1,782	2,164	3,022	4,414	6,175
Settlements	2,450	2,353	2,301	2,666	2,733
TOTAL	4,232	4,517	5,323	7,080	8,908
Students beyond compulsory school age	96	165	287	560	1,069
Kindergarten children					525
TOTAL	4,328	4,682	5,610	7,640	10,502

(*Source: Development of the educational system in Greenland 1950–70* (undated), by C. Berthelsen.)

tional training. In accordance with the then prevalent philosophy of integration, and to meet the wishes of the Greenlanders, who wanted Greenland to be brought into the modern world without delay, much greater emphasis was placed on the teaching of the Danish language and more advanced education, with a large number of students going to Denmark for specialized training.

The authorities were faced not only with improving the quality and variety of instruction, but also with a rapidly increasing number of students. By 1970 there were two-and-a-half times more pupils in Greenland than there had been in 1950, with the increase almost completely in the towns as a result of the urbanization that was taking place (Table 27). In Godthaab the number of pupils increased by a factor of six in these twenty years. The additional teachers needed had to come from Denmark and could not speak Greenlandic. Danish-speaking teachers, who in 1950 had formed only 15 per cent of the total number of teachers, accounted for 71 per cent by 1970, and their numbers had risen from 33 to 401 (Table 28). The effort put into education has begun to bear academic fruit and the number of pupils passing the school-leaving examination is rising rapidly, while their average age at this time is much lower. At the same time there has been growing concern that the Greenlandic culture is being eroded, leading to a reaction against the emphasis on the Danish language in the school system.

Great efforts are also being put into vocational training to prepare the Greenlanders for an urbanized and technical society. The policy is that all training programs should as far as possible be carried out in Greenland. When, however, the training is of a nature that cannot be given in Greenland, the trainees are sent to Denmark. In 1974 there were over 1,000 Greenlanders taking vocational courses of a year or more in length in either

Table 28. Number of Greenlandic-speaking and Danish-speaking teachers, 1950–70

	1950	1955	1960	1965	1970
Greenlandic-speaking	182	184	163	160	165
Danish-speaking	33	68	94	198	401
TOTAL	215	252	257	358	566

(*Source*: *Development of the educational system in Greenland 1950–70* (undated), by C. Berthelsen.)

Table 29. Number of Greenlanders taking vocational training of one year or more in 1974

	Greenland	Denmark	Total
University or equivalent	—	15	15
Teacher training	117	11	128
Other pedagogic, social welfare and public health training	125	73	198
Technical training	—	26	26
Navigation training	—	5	5
Training of apprentices in			
Construction	76	51	127
Graphics	5	—	5
Office and business occupations	239	9	248
Iron and metal trades	130	60	190
Food industries	24	12	36
Service industries	6	21	27
Other training programs	—	35	35
TOTAL	722	318	1,040

(*Source*: Adapted from Sven Koch, *Vocational training in Greenland*, Copenhagen, 1975, p. 12.)

Greenland or Denmark (Table 29), not counting about 100 training in nautical schools and 90 attending folk high schools in Denmark.

The need for more highly educated Greenlanders will become more pressing with the advent of Home Rule and the policy of Greenlandicization that will inevitably accompany it. The fall in the birth-rate will mean that less emphasis has to be given to building new facilities, but the supply of teachers will be a very difficult problem. Danish teachers are finding the terms of employment less attractive, but improvement would run counter to the policy of equalizing the treatment of Greenlanders and Danes. A teachers' college has been established at Godthaab to provide a two-year training course for native teachers who do not have the qualifications for the regular four-year college course, but it will be able to meet only a small part of the need. Until the number of well-educated Greenlanders is greatly increased, Greenland will remain dependent on Denmark not only for many of her teachers, but also for filling most of the higher administrative and specialist positions, even if development of the country and its resources proceeds more slowly. Unfortunately the drop-out rate of Greenlanders attending university in Copenhagen is very high.

Health services have expanded greatly since the end of the Second World War. Their first major task was an attack on tuberculosis which was widespread. In 1952 some 7 per cent of the population were suffering from active tuberculosis which was the cause of 40 per cent of all deaths. A vigorous, sustained campaign and the great improvement in the standards of housing have been so successful that tuberculosis is now responsible for only one or two deaths annually. The numbers of doctors, nurses, and hospital beds have increased even more rapidly than the population and the number of people per doctor is now much the same as in Denmark. The effectiveness of such health measures can be seen from the statistics. The death-rate fell from 24.1 per thousand population in 1950 to 8.3 in 1967, admittedly a rather deceptive figure because of the large number of young people in the population. In the period 1946–51 the number of children dying in their first year was 172 per thousand; for the period 1966–70 the figure had fallen to 57. The expectation of life at one year of age in 1946–51 was 38.5 years for males and 43.4 for females; in 1966–70 it was 61.9 for males and 68.4 for females. Accidents are now responsible for over one-quarter of all deaths, the majority from drowning.

Epidemic diseases occur frequently but the population has acquired a measure of immunity and the effects are much less serious than they had been previously. Dental decay, which had been very rare when hunting was the main occupation, has become widespread with the change to a diet rich in carbohydrates. A cause of great concern is the increase in venereal diseases which are very difficult to control in the conditions prevailing in Greenland. In 1973, the reported cases of gonorrhoea numbered 11,721, many of them reinfections, and of syphilis 246. Gonorrhoea has been easy to cure, but penicillin-resistant strains are beginning to appear.

The use of alcohol is another social problem of major dimensions in Greenland, and it is not restricted to the native population. In the year 1968–9 the expenditure on beer, wines, and spirits in the average native household (5.1 persons) was 2,000 kroner out of a cash income of 26,518, while for the average Danish household (3 persons) it was 3,200 kroner out of a cash income of 59,495.

The physical conditions under which the Greenlanders live have improved very greatly during the past twenty years. Housing is of a much higher standard, and incomes, though far below those in the rest of Denmark, have risen rapidly. In the ten years from 1955 to 1965, the *per capita* real income of the Greenlanders doubled and the rise has been sustained since. These changes have, however, been accompanied by a number of serious social problems. The many immigrants to the towns have been faced with conditions completely different from those to which they had been accustomed in the villages they had left, and have found it difficult to adjust to so much that is strange. The informal but powerful social controls that regulated life in a village are no longer present and family ties have been loosened. The towns offer many distractions, and the immigrant finds himself in situations to which he does not know how to respond. The gap between the more adaptable youth and their more conservative parents has widened still farther. To many, alcohol provides a refuge, though not an answer. The very high incidence of venereal disease is an indication of the social problems, and about half the births are illegitimate. Another disturbing sign is the increase in mental illness, which was rarely present or little recognized in the past.

An indication of the social malaise is the increase in crime. Thefts have become commonplace, but it is the growing numbers of murders and other crimes of violence that is most disturbing. Violence is not restricted to personal quarrels among Greenlanders, and in recent years a number of Danes have been assaulted. This hostility may be deep and of long-standing but, if so, it has never come to the surface before. Relationships between the Danes and the Greenlanders in the past always appeared to be based on mutual respect and tolerance. Animosity, though inevitably present to some degree, rarely prevailed, mostly owing to the Greenlanders' traditional respect for authority. One factor in the change is certainly the great increase in the number of Danes on the island and their continued pre-emption of almost all the positions of power. Danes now going to Greenland are not as carefully selected as in the past and they are there for different reasons. Most intend to stay for a comparatively short period in order to save up money. They do not have a long-term commitment or a lasting interest in the country, and there are so many of them that they can form their own exclusive society. In Godthaab and other towns which have been developing rapidly, Danish construction workers, with money to spend, are in competition with Greenlanders not only for jobs but also for women. Table 30 gives the incomes and expenditures of a sample of Greenland households in 1968–9. It shows the large incomes of those who are sent out from Denmark, more than twice those of the indigenous people, while their families are much smaller. Despite the far higher standard of living they enjoy, they are able to save nearly a quarter of their income. The Greenlanders, in contrast, can only just make ends meet.

Table 30. Expenditures and incomes of households of wage and salary earners in Greenland, 1968–9

	Indigenous				Sent out from Denmark	
	Salaried employees and public servants		Wage-earners			
Number of households	125		241		140	
Number of persons per household	4.9		5.2		3.0	
Expenditure on:	kr.	%	kr.	%	kr.	%
Dwellings	1,442	4.9	1,472	5.9	485	0.8
Fuel and lighting	2,041	6.9	1,623	6.5	2,070	3.5
Food	7,834	26.6	7,479	29.8	8,798	14.7
Beverages	2,172	7.4	1,907	7.6	3,285	5.5
Consumption outside the home	356	1.2	315	1.3	1,073	1.8
Tobacco	2,266	7.7	2,258	9.0	2,474	4.1
Clothing	3,025	10.3	2,740	10.9	3,674	6.1
Footwear	746	2.5	643	2.6	681	1.1
Washing and cleaning	380	1.3	361	1.4	593	1.0
Furniture and household utensils	2,038	6.9	1,434	5.7	4,034	6.7
Personal hygiene, nursing	558	1.9	443	1.8	983	1.6
Books, newspapers, telephone, radio, etc.	1,131	3.8	805	3.2	3,248	5.4
Holidays in Greenland	280	0.9	161	0.6	307	0.5
Holidays in Denmark	383	1.3	99	0.4	2,636	4.4
Holidays abroad	24	0.1	2	0.0	538	0.9
Other leisure activities	269	0.9	172	0.7	1,276	2.1
Transport	373	1.3	282	1.1	1,381	2.3
Miscellaneous	2,474	8.4	1,700	6.8	6,027	10.0
Hunting, fishing, etc.	1,186	4.0	1,198	4.8	1,740	2.9
Total consumption	28,978	98.4	25,094	100.1	45,303	75.5
Saving	483	1.6	−27	−0.1	14,678	24.5
Consumption+saving	29,461	100.0	25,067	100.0	59,981	100.0
Cash income	29,399		24,991		59,495	
Income in kind	961		1,099		213	

(Source: Statistisk Årbog, 1976, Copenhagen, Table 458.)

In this sort of a situation, racial relations cannot avoid becoming strained. The great majority of the Greenlanders, however, especially among those who are better educated, realize that Greenland owes a great deal to Denmark and, while they may have reservations about some of the policies and attitudes that the Danes have adopted, they have no doubt about the humanity and sincerity of their intentions. At the same time most of the Danes in Greenland understand the nature and severity of the problems that are facing the Greenlanders today, and recognize that Denmark must bear responsibility for the few failures as well as for the successes that have resulted from her policies.

Trade

For 175 years the Royal Greenland Trading Company enjoyed absolute control of trade in Greenland by buying and marketing all products, purchasing and retailing to the Greenlanders all their supplies, controlling transportation to and from the island and along the coast, establishing and operating fish-processing plants, and conducting any other commercial activity that might affect the economy. It was protected from any competition and, as part of the administration, it was assured of full government support. The Hudson's Bay Company, always attempting to achieve a similar position in northern Canada, must have looked on the Royal Greenland Trading Company with envy.

When the monopoly was ended in 1950, it was expected that private Danish merchants and industrialists would move in quickly to establish commercial undertakings. There was concern that private enterprises would attempt to maximize profits by charging as much as they could and paying as little as possible, while avoiding anything that did not seem likely to produce a good return. Government controls were established to prevent this from getting out of hand. It was feared that, the cream having been skimmed off the trade, the Royal Greenland Trading Company would be left with the responsibility of supplying the smaller places and continuing only those operations that were unprofitable. In fact, however, no rush developed. Danish commercial interests have been hesitant about moving into Greenland. A number of private businesses have been established, especially in the open-water ports and sometimes with substantial government assistance, but the Royal Greenland Trading Company has continued to dominate the trade, and recently it was again given control over the shipment of supplies between Denmark and Greenland. Private businesses are almost all controlled by Danes, as the Greenlanders, living so long in a non-competitive economy, do not have the experience to compete in the business world. Many of these Danes marry Greenlanders and settle permanently in the country. The Royal Greenland Trading Company is now being cast in the role of custodian of enterprises until they can be taken over by Greenlanders. The Greenlanders themselves have mixed feelings about the Royal Greenland Trading Company. They frequently consider it slow-moving, old-fashioned, and out of touch with reality, criticisms that government undertakings of this nature attract, often unjustly, the world over, but they recognize the valuable service it performs and they appreciate the fact that nearly 90 per cent of its employees in Greenland are Greenlanders. They contrast this with the Greenland Technical Organization, which is of course faced with a very different task, and in which many more employees are Danes.

A development of considerable interest is the establishment of co-operatives in a number of towns. In some cases they have taken over the Royal Greenland Trading Company's stores while in others they operate in competition with them as well as with private traders. The movement seems to be growing and by 1974 seven co-operative stores were operating and five others were being organized. The principle of transferring retail stores from the Royal Greenland Trading Company to co-operatives has been endorsed by the Greenland Provincial Council, and is at least part way along the road leading to the administration's objective of encouraging private enterprise, but the political and economic implications are being carefully examined before definite policies and procedures are adopted.

Government and administration

Greenland has been constitutionally part of the Kingdom of Denmark since 1953 and elects two members of the Danish Parliament in Copenhagen. At first one represented the west coast from Holsteinsborg to the south, and the other the west coast north of Holsteinsborg and East Greenland, but in 1975 the whole island became one constituency. On the basis of population, Greenland's representation of two members is very generous compared with the rest of Denmark. As the Danish Parliament has only 179 members divided among eleven political parties the support of one or two members can be crucial to a government. The Greenland members can therefore be, and have been, in positions of great political power, and two Greenland members have been Minister for Greenland.

The Greenland members were at first independents, elected on an individual basis and owing allegiance to no political party. In 1975 however a political party emerged in Greenland, the Siumut, with nationalistic left-oriented leanings, to which one of the Greenland members belonged. In the 1977 elections there was a second Greenland political party, Atassut, which was also committed to Home Rule but favoured close co-operation with Denmark and was more moderate in politics. Voting was fairly evenly divided, one seat going to each of the two parties. The Greenland members may remain independent in the Danish Parliament, but sometimes they align themselves with existing Danish political parties.

The Greenland Provincial Council (Grønlands Landsråd), which has its seat in Godthaab, at first covered only West Greenland, but North Greenland and East Greenland have been included since 1961 and it now represents all Greenland. There are sixteen constituencies, thirteen in West Greenland, two in East Greenland, and one in North Greenland.

Each elects a member, and anybody domiciled in Greenland is entitled to vote provided he is eligible to vote in the Danish Parliamentary elections. The number of members can be increased to secure suitable representation of divergent views and interests and the Council elected in April 1975 included a representative of the Greenland Workers' Union. The responsibilities of the Council are partly advisory and partly executive. All proposed measures exclusively concerned with Greenland must be submitted to the Greenland Provincial Council to express its opinion before they are introduced in the Danish Parliament. Drafts of administrative regulations and decisions affecting the people of Greenland are also referred to it, and it can put forward proposals to the government and submit questions and complaints concerning any Greenland affairs that do not come under its own jurisdiction. The Provincial Council is itself responsible for social welfare, for drawing up the regulations covering the awarding of social grants, and for the protection of wildlife. School education and health are under the jurisdiction of the Danish state.

At first the elected members of the Greenland Provincial Council came very largely from those who had been state employees, but after the elections of April 1975 other interests were better represented and now provide about half the members. The Council derives its revenues from both indirect taxation, especially on luxury items such as alcohol and tobacco, and direct taxation. There is also a state grant of 30 per cent of the costs of social welfare. The Council appoints an Executive Committee to deal with matters requiring attention when it is not in session.

At the local level Greenland is divided into nineteen districts, each with a town which acts as the centre for a number of small settlements. The districts have their own elected councils, responsible for welfare, police regulations, development planning, garbage collection, allocation of dwellings built with the aid of state subsidies, and certain other matters, and they are financed by annual subsidies provided by the state and the Provincial Council, and by municipal income taxes.

The Ministry for Greenland is responsible for health, education, public works, housing, trade and industry, and many other activities, and includes the Royal Greenland Trading Company and the Greenland Technical Organization. Two scientific institutes, the Geological Survey of Greenland and the Greenland Fisheries Survey, also form part of the Ministry. Certain responsibilities that used to be included within the Ministry for Greenland now fall under other Danish departments. For example, the Greenland Police, as part of the Danish State Police, are the responsibility of the Ministry of Justice, the Church comes under the Ministry of Ecclesiastical Affairs, while broadcasting and libraries went to the Ministry for Cultural Affairs in 1967 but returned to the Ministry for Greenland in 1972.

The Minister for Greenland is represented in Greenland by the Governor, who is the supreme administrative authority in the island. He does not however have control over the Royal Greenland Trading Company or the Greenland Technical Organization, both of which are managed directly from Copenhagen. Until 1967 the Governor chaired the Provincial Council but it now elects its own chairman and has developed its own secretariat. The Governor has the right to take part in debates, presenting the government point of view on any matter and offering help and advice, but he does not vote. He is also a member of the Welfare Board chaired by the Chairman of the Provincial Council. The Provincial Council on the other hand nominates members of the Board of Education, the Housing Loan Fund Board, the Radio Broadcasting Board, and other important agencies of the administration. In ways like these the Provincial Council and the Greenland administration form two distinct but interlocking hierarchies.

In 1964 the Danish government established the Greenland Advisory Council in Copenhagen to watch over the development of the island and to make recommendations to the Ministry for Greenland, particularly regarding the planning and co-ordination of public works. It consists of a chairman appointed by the Crown, the two Greenland Members of Parliament, a member from each of the five largest parties in Parliament, and three members elected by the Greenland Provincial Council. The Greenland Advisory Council provides a forum where Danish and Greenland politicians can freely discuss Greenlandic problems at an early stage. It is purely advisory but its advice is very influential, especially when such matters are being considered as the priorities for long-term government investments in the development of Greenland.

Home rule

The Home Rule Commission has held a number of meetings in both Denmark and Greenland. Agreement has been reached on many issues and the general pattern of Home Rule is taking shape. Some matters, however, are apparently proving difficult to resolve, and there will be little time for public discussions, a referendum, and the completion of legislation by the Danish Parliament before 1 May 1979, when Home Rule is to begin.

Complete independence, as in Iceland, does not appear to be the goal of either Denmark or Greenland. It is rather the Faeroe Islands model that will be followed, with Greenland remaining an integral part of Denmark, and the Danish Parliament delegating its responsibilities for Greenlandic matters to a Greenlandic authority. The Danish Parliament will retain responsibility for matters affecting the entire nation. These include the constitution, foreign policy, defence, national finances, and the legal system, but even in these areas special laws and regulations will be possible to meet the special conditions and interests of Greenland, as is the case today for instance with the special Greenland Criminal Code. Greenlandic will be the first language but Danish will also be used in public affairs. The Greenlandic Church will be part of the Home Rule administration. The Greenland Provincial Council will become the Greenland Legislative Assembly, giving up its advisory role and assuming responsibility for legislation in those fields delegated to Home Rule. A small Greenland Council will act as a Greenland cabinet, its members residing in Godthaab and each administering some area of government operations, such as education, social affairs, cultural affairs, and finance. The Council will be empowered to act only within guidelines determined by the Legislature, which will decide the budgetary limits in each area that the Council member concerned cannot exceed. Co-ordination and responsibility for the government system will rest with the Chairman of the Council, whose office will prepare the agenda for meetings of both the Council and the Legislature. The Chairman and members of the Council, who need not be members of the Legislative Assembly, will be appointed for the same period for which the Legislature is elected.

In theory the financial arrangements should be for Greenland to pay for all matters for which she assumes full legal authority and administrative responsibility, for Denmark for the time being to pay for all matters for which she retains full responsibility, and for mixed financing for matters that fall in between. In practice, however, it is admitted that Greenland is not in a position to finance the operations that will become her responsibility under Home Rule, and that Denmark will have to continue, for many years at least, to subsidize Greenland at a level comparable with current practice. This will be done by transferring block sums, with the Greenland Legislative Assembly determining how this money will be allotted. It will be an even longer time before Greenland can be expected to contribute towards the costs of matters of national concern such as defence.

There is an air of unrealism about any form of Home Rule that depends on large and continuing subsidies, and is epitomized by the proverbs 'Beggars cannot be choosers' and 'He who pays the piper calls the tune'. The principle of Home Rule is to provide the country with the freedom to choose its own course. Different courses involve different levels of expenditures and, unless the country generates its own revenues, this entails asking for money. The only way in which Greenland sees a possibility of escaping from this dilemma is through the exploitation of mineral resources. An issue of the greatest importance to Home Rule therefore is the ownership of sub-surface minerals. The first paragraph of the 'Act on Mineral Resources in Greenland' stipulates that all mineral resources in Greenland belong to the state. This reserves to the Danish state the right to issue licences to exploit the mineral resources. The Act later refers to the economic interest of the state and of Greenland's provincial treasury being secured by royalties. Both the Home Rule Commission and the Greenland Provincial Council have agreed that the first paragraph of the act will have to be amended.

The revenue from minerals over the next few years will be very small in relation to Danish expenditures in Greenland, and it might appear to make little difference whether royalties were paid to the state or to the Greenland treasury where they could offset part of the state subsidy. There is however the possibility, and some have even considered it a probability, that offshore oil will be discovered in substantial quantities. Extraction would have to be on a very large scale in order to be economic, and the annual amount accruing to the government from this source, now zero, could rise to several billion kroner and far exceed the annual government grants. It is a feast or famine situation. The Danish government, having made such large expenditures in the post-war years in attempting to raise the standard of living in Greenland closer to Danish levels, is naturally reluctant to agree to provisions that might, however improbably, lead to rich Greenlanders making takeover bids for Danish industry and mutterings in the streets of Copenhagen about Greenland sheiks. Some compromise will be necessary. This will probably be based on equal status between Denmark and Greenland concerning the administration, policy, and management of raw materials, and a joint council, with equal representation from the Danish government and the Home Rule administration, making recommendations to the two parties. The Home Rule Commission is of the opinion that Greenland's share of any oil income must at least equal the present expenses of the Greenland treasury.

Membership in the European Economic Community was an issue on which the views held in Denmark and Greenland differed widely. Denmark could see significant advantages in becoming part of a larger economic unit. Greenland viewed the European Common Market as a grave threat to her fishing industry, and a way in which resources of the island could be exploited for the benefit of European industry, with Greenland becoming the hinterland to a continent. Free access to Greenland could quickly lead to the Greenland economy and labour market being completely dominated by outsiders. Having spent over two centuries as a colony of Denmark, Greenland did not wish to become a *de facto* colony of Europe. She joined the European Economic Community against her wishes, but under Home Rule she will be able to decide whether or not to withdraw. The European Economic Community will not want to lose Greenland and will make every effort to meet her special needs and to demonstrate the advantages of membership. The European Industrial Bank has made loans of over 90 million kroner to Denmark to finance such undertakings as a new power station at Godthaab, improvements to several harbours, and a telecommunications network between Kap Farvel and Disko Bugt. In 1975 the European Economic Community's regional fund contributed 25 million kroner to various projects in Greenland, and its social fund provided 20 million kroner towards technical education and vocational training. There is however a belief in Greenland that these sums have been offset by reductions in the subsidies provided by Denmark, so that Denmark rather than Greenland has been the beneficiary. The question of membership in the European Economic Community will be the first great debate in Greenland after Home Rule, and much will depend on the result. At present the odds are that Greenland will decide to leave.

Transport

The sea has always provided the highway not only between Greenland and Denmark but also within Greenland. The population lives along the coast which is usually steep and rocky and often precipitous, making communication by land impracticable. Contact was maintained between the settlements by sea either in boats or, when the sea was frozen, by dog sled. Overland travel has been used only for scientific expeditions on the ice cap and for purely local purposes within towns.

The pattern of transport that has developed in West Greenland is for Julianehaab to receive direct shipments of all classes of goods. Large bulk consignments for other places are usually sent to the 'trans-shipment harbours' of Godthaab and Egedesminde, and distributed from there by coastal shipping directly to all settlements in southern Greenland and northern Greenland respectively. Other goods are shipped to the major towns in each district and then distributed to the smaller places in their districts. However, with improved storage facilities and with the great increase in freight, which is now over 100,000 tonnes a year, it has become more economical to send some bulk consignments, such as petroleum, directly to towns rather than through the trans-shipment harbours.

The major change in transport since the Second World War has been the introduction and development of travel by air. Before the war the only aircraft that had been seen in Greenland were a few pioneer flights, but during the war military airfields were built by the United States at Narssarssuaq, Søndre Strømfjord, and near Angmagssalik, and large numbers of aircraft passed through them. A few years after the war another major military airfield was constructed at Thule in North Greenland. The post-war development of long-range civil aircraft made commercial transpolar flights possible, and Scandinavian Airlines System soon began to use Søndre Strømfjord as a refuelling point on flights between Europe and America and to carry passengers between Denmark and Greenland. The journey by ship is both time-consuming and, in the winter months, dangerous, and within a few years the air had captured virtually all the rapidly increasing passenger traffic to and from Greenland. In 1953 passengers arriving or leaving by ship totalled 5,439, and by aircraft 1,121. By 1973 the number of passengers by sea had fallen to 643, and those by air had risen to over 34,046.

Greenland looked also to the air for improved internal transport but was faced with the problem of finding airfields in a country where level ground is almost non-existent. On the initiative of the Royal Greenland Trading Company, Eastern Provincial Airways, a Canadian company, began a service in 1958 along the west coast with amphibious aircraft. Two years later Greenland Air was formed by Scandinavian Airlines System and the 'Oresunde' cryolite company to serve the United States military bases. In 1962 Greenland Air took over the Eastern Provincial Airways operation in Greenland, and both the Greenland Provincial Council and the Royal Greenland Trading Company took shares, so that 25 per cent of the shares are now owned by each of the four parties. In 1965 Greenland Air began to operate a helicopter service linking the towns along the west coast. Helicopters are, however, very expensive to operate and maintain and require good flying conditions. The service, though certainly filling a real need, has always been very costly, there have been long and frequent delays owing to weather conditions and unserviceability, and one or two serious accidents have occurred. Greenland Air would not be able to operate without a subsidy, which is now about 175 kroner for each passenger carried. Short take-off and landing fixed-wing aircraft could provide a faster, more reliable and less expensive service, can fly on instruments in poor weather and require only short airfields with much steeper angles of approach than are necessary with conventional aircraft, and it would be possible to locate these near to many of the twenty-one places on the west coast now served by helicopters. It seems likely that short take-off and landing aircraft will soon begin to replace the twelve helicopters now owned by Greenland Air, and the Greenland Provincial Council has recommended the construction of a 900-metre airstrip near Godthaab to start in 1977. Some helicopters will however continue to be needed to serve places where even a short airstrip is impracticable as well as for rescue operations.

Greenland Air has captured part of the internal passenger traffic but it does not serve every community and many people still prefer to travel by sea. The Royal Greenland Trading Company operates two coastal ships which carry passengers on the west coast between Nanortalik and Upernavik from May to November, as well as several smaller ships, and a third ship operating between Søndre Strømfjord and Godthaab. The passenger service by sea is no more profitable than that by air. In 1973 it operated with a deficit of 11 million kroner which compares with a deficit of 9 million kroner for Greenland Air in 1974.

Foreign transpolar flights using Søndre Strømfjord for refuelling are not permitted to carry passengers to and from Greenland, which has external air services to Denmark and Iceland only. Proposals to establish a service between North America and Greenland have received little encouragement in Denmark which has reason to be apprehensive of increased North American influence in Greenland. North America could easily replace Denmark as the source of supply for many commodities and it is a powerful magnet in other ways. There is, however, increasing interest among the Greenlanders in their kinsmen to the west and it can only be a question of time before a regular service is begun between Greenland and North America. It would certainly follow hard on the heels of Home Rule. In the meanwhile a Greenlander can look across Davis Strait towards North America but a visit there would usually involve crossing the North Atlantic four times.

Dogs are still widely used north of the 'dog line' at Holsteinsborg, south of which it is illegal to keep sled dogs. Unlike other northern areas the motorized toboggan has not made many inroads and its use for hunting is discouraged.

The future

The post-war years have seen enormous expenditures by the Danish state with respect to Greenland. It was hoped that with this capital investment Greenland would develop an economy in which private Danish capital and enterprise could operate freely and which would eventually become self-supporting. It would also provide the way for raising the standard of living in Greenland closer to the standard prevalent in Denmark.

Danish state investment in Greenland rose rapidly from 12 million kroner in 1951 to 280 million kroner in 1968, and operating expenditures also increased. In twenty years from 1951 to 1970 state expenditures totalled about 4,500 million kroner and by the end of this period were at a rate equivalent to 15,000 kroner annually for each Greenlander. Not all of this money was spent in Greenland and much of it went to purchase Danish supplies and services, but the monetary income of Greenlanders nearly tripled between 1955 and 1970. At present, capital transfers from Denmark amount to about 1,200 million kroner annually. Expenditures have decreased a little recently partly owing to the slower rate at which the population is increasing and partly reflecting the wishes of the Greenland Provincial Council to temper the rate of growth.

The increase in government expenditures has been accompanied by a rapid increase in imports to Greenland and a much slower growth of

Table 31. Imports to Greenland – 1974 (thousand kroner)

Meat and meat preparations	24,744
Dairy products and eggs	10,483
Fish and fish preparations	2,370
Cereals and cereal preparations	10,202
Fruits and vegetables	18,394
Sugar, sugar preparations, honey	10,814
Coffee, tea, cocoa, spices	7,624
Feeding stuffs for animals	4,110
Miscellaneous food preparations	8,850
Beverages	28,323
Tobacco	6,254
Wood, lumber, cork	15,173
Coke and coal	3,025
Petroleum and petroleum products	99,468
Chemicals and chemical products	4,273
Material for tanning and dyeing	4,336
Medical and pharmacology products	7,134
Vegetable oils and cosmetics	7,847
Explosives	3,085
Plastics	3,141
Rubber manufactures	2,680
Wood manufactures	36,676
Paper, paperboard manufactures	16,918
Textiles	19,483
Non-metallic mineral manufactures	10,737
Metals	16,450
Manufactures of metal	37,789
Machinery other than electric	54,907
Electric machinery	49,742
Transportation equipment	20,123
Sanitary, heating and lighting fixtures	6,512
Clothing and footwear	27,744
Instruments, watches, photographic equipment	8,338
Miscellaneous manufactures	27,312
Other imports	21,570
TOTAL IMPORTS	633,691

Apart from the coal, which came from Poland, and the petroleum, most of which was Venezuelan, virtually all imports were from Denmark, though not necessarily of Danish origin.
(*Source: Statistisk Årbog, 1976*, Copenhagen, Table 449.)

exports. In 1954 imports at 71 million kroner exceeded exports by 41 million kroner, by 1970 the annual deficit had reached 289 million, and in 1973 it was 375 million. The next year however saw a dramatic improvement and the deficit fell to about 80 million owing to exports of lead and zinc concentrates from the 'Black Angel' mine (Tables 31 and 32).

Many of the factors that will determine the course of the economy of Greenland are unpredictable. Will the cod return in abundance and restore the base on which the Greenland fishing industry and the plans for

Table 32. Exports from Greenland, 1974 (thousand kroner)

Meat of sheep and goats	667
Other meat	1,886
Fish, fresh, chilled or frozen	74,462
Fish, salted, dried or smoked	10,693
Crustaceans and molluscs	5,835
Fish preparations	96,636
Meat meal and fish meal	3,200
Fur skins, undressed	9,490
Quartz, mica, feldspar, cryolite	6,355
Lead ore and concentrates	66,806
Zinc ore and concentrates	269,948
Bones, ivory, horns, claws	3,134
Other categories	1,982
TOTAL EXPORTS	551,094

Most exports were to Denmark, but fish to the value of about 36 million kroner was exported to the USA, and the metallic concentrates went mainly to France (over 2/5), Finland (under 2/5), and the Federal Republic of Germany (1/5).
(*Source*: *Statistisk Årbog, 1976*, Copenhagen, Table 450.)

modernizing the country had been built? Will minerals, and in particular petroleum, be found in quantities that will make Greenland rich, if only for a limited period? These are questions that time, rather than the Greenlanders, will answer. One question that only the Greenlanders can answer is whether they will remain in the European Economic Community. If they elect to leave, it must inevitably lead to some weakening of their close economic ties with Denmark. How will this be adjusted? Will Greenland then turn her eyes east and form closer links with those other North Atlantic countries, Norway, Iceland, and the Faeroe Islands, which also have important fishing industries and which have chosen to remain outside the European Economic Community? Both Norway and Iceland at one time formed part of the Danish kingdom, and the Faeroe Islands have remained within it. Will Greenland turn rather to the west towards North America, whence she was originally populated, where she shares a common cultural heritage with the Canadian Eskimo, and where markets can be developed for her products? In any event, the close relationship between Denmark and her former colony, now part of her kingdom, rests on more than economic considerations and will remain strong, no matter what economic changes take place.

CHAPTER SIX

The north of the Old World: lands of the northeast Atlantic

Northern Scandinavia

The whole of the Scandinavian peninsula may appear northern to both North Americans and Russians, for it lies farther north than Edmonton or the Trans-Siberian railway. But mere latitude is a bad guide. The warming influence of the Atlantic is strongly felt, and the only part of the peninsula which properly concerns us is the so-called 'northern cap' or Nordkalotte, named from its similarity on the map to a priest's cap (*kalotte*). The term is not closely defined, but one may take it to comprise (and many Scandinavians would agree) a tract of land rather bigger than that limited by the Arctic Circle, consisting of the most northerly counties of the three countries concerned: Finnmark, Troms and Nordland in Norway (collectively known as the province of Nord Norge), Norrbotten in Sweden, and Lappi in Finland (Map 17).

Although there are marked differences between the component parts of the northern cap – geographical, political, economic, and cultural – the events of the last quarter of a century make it logical to treat this as a single but diversified unit. The Scandinavian countries have been drawing steadily closer to each other, and nowhere is this more marked than in the north, where they also meet physically.

Geographical and historical background

This is the area of the Baltic or Fennoscandian shield, a stable platform of Precambrian rock. Folding occurred on its western edge in the Silurian

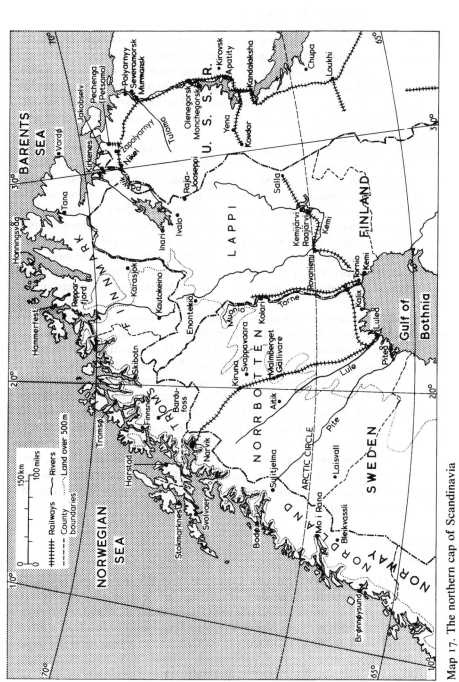

Map 17. The northern cap of Scandinavia

period, giving rise to the fjorded coast on the western and northern flank, and coastal mountains with Alpine relief and elevations up to 2,000 m. Inland, the mountains give way to a plateau, which in turn drops away to the southeast and the shield proper. The whole of the area was heavily glaciated under an ice sheet which stretched to the British Isles in one direction and Siberia in the other. But today there is little ice cover and little permafrost in the area.

The climate varies considerably across the region. The Atlantic coast is the mildest land area in the world at this latitude. The sea does not freeze, but snowfall can be heavy, and the sun is absent for two months in the winter. Over the mountains, to the east and south, much lower winter temperatures are found. The head of the shallow Gulf of Bothnia is ice-covered for five months, and the climate begins to become continental. The soil is rather poor, and supports mainly forest vegetation, but the uplands are altogether treeless. The water resources of the area are considerable. Short, precipitous streams empty into the ocean on the west and north; longer, slower rivers flow into the Baltic; and there are many lakes.

The northern cap is by European standards remote, wild, and harsh, but by comparison with the other territories considered in this book it is close to major centres in the south and by no means extreme in climate. By any standard it is handsome country, and often spectacularly beautiful.

The Roman historian Tacitus knew about northern Scandinavia in the first century AD. The first circumstantial account by a local traveller is that of the Norseman Ochtere or Ottar, who lived near modern Tromsø and recounted his story to King Alfred of England in the ninth century. Thereafter, many ships from Western Europe reached the area, which was incorporated into the Kingdom of Norway at about the same time. The internal frontiers were defined in the late eighteenth and early nineteenth centuries, but important changes were made to the Finnish–Russian boundary in 1920 and 1944.

Resources

Minerals

Iron is the dominant mineral now exploited, and Kiruna in Norrbotten is the major mining centre. The deposit of high-grade iron ore found there yields 17–19 million tonnes of 60 per cent iron a year (average for 1971–5). Neighbouring Gällivare and Malmberget (where iron-mining first started in the eighteenth century) produce 5–7 million tonnes a year, and the most recently opened mine, at Svappavaara, yields 2–3 million tonnes. Norrbotten iron is the major extractive industry in Scandinavia. Most of the product is exported through Narvik, and about a quarter through Luleå at the other end of the railway. In Norway, Finnmark has the country's largest iron mine at Sør-Varanger (Kirkenes), where in 1975 the low-grade ore yielded 2.3 million tonnes of pellets from an opencast pit located 5 km from the Soviet border. Some ore for this plant is imported from the Murmansk region. Finland obtains iron ore, too, in small quantities, at Kolari (45,000 tonnes of concentrates in 1975). Another mine, Raajärvi, closed in 1975, but it is intended to increase output at Kolari to 0.5 million tonnes of concentrates a year.

There are some exploitations of other minerals: copper at Sulitjelma and Repparfjord, lead and zinc at Bleikvassli and nearby Mofjellet, all in Norway. The lead mine at Laisvall in Sweden is the biggest in Europe, producing over a million tonnes of 4 per cent lead ore a year. There is some opencast mining of copper at Aitik in Sweden, and a chrome mine at Kemi in Finland. A valuable resource was the nickel deposit at Petsamo, at which development started in 1934, but this was surrendered to the USSR in 1944. It was no doubt one of the reasons why the USSR wanted the territory.

The presence and exploitation of apparently heavier mineralization across the Soviet border (nickel, iron, copper, aluminium, apatite) leads Scandinavian prospectors to suppose that more discoveries may be made in the northern cap.

Forests

There are large forested areas in that part of the northern cap south and east of the coastal mountains. At this latitude the trees grow slowly, but although they take twice as long to mature as they do in southern Scandinavia, there are nevertheless exploitable reserves of pine, spruce, and birch. These are owned both by the state and by individual farmers. There are major softwood-processing plants at Kalix and Piteå in Sweden, and at Kemi and Kemijärvi in Finland. There have been changes in the manner of operation of the industry over recent years: mechanization of many processes, reduced use of rivers, increased use of trucking and rail transport to take timber to selected automated ports. In Finland, there is controversy over whether northern timber should be used only in the north or exported to the south. Although these northern enterprises produce much less than their counterparts farther south, their output may grow, for forest industries are likely to increase, but at the expense of farming. With careful management the industry may have a good future.

Farming

The remoteness of the area, before the transport improvements of the last hundred years, favoured the prevalence of subsistence farming. This was generally animal husbandry, and the long winter meant that fodder crops had to be grown and stored. But plant-breeding experiments of the last half-century have made possible a northern advance of cereal crops, and these are now found at latitude 68° N in the Muonio valley, for instance. However, the technical possibility does not necessarily imply economic expediency, and in fact arable farming does not flourish in the north. Dairy farming continues, however, and may be found in favoured locations at even higher latitudes – the Pasvik valley, Tana, and Alta. Likewise potatoes can be made to grow at many sheltered spots very far north. But the improvements that are now possible in the north are being applied also in the south, so southern Scandinavian farmers are able to reduce their opposite numbers in the northern cap to a very marginal position – and there is, in addition, an overall excess of farm products in Scandinavia.

One branch of farming deserves special mention – reindeer husbandry. Most reindeer live within the northern cap, and by their choice of food, preferentially lichens of the *Cladonia* family, compete with no other animal. They provide not only meat but hide, fur, milk, antlers for crafts, traction, and pack transport, so they are in many respects the ideal basis for northern animal husbandry. There are about 600,000 domesticated reindeer in northern Scandinavia (1974). Traditionally they have been herded by Lapps, and in Sweden only persons of Lappish descent may own herds. About 2,000 Lapps are now directly involved in herding, and perhaps another 8,000 persons are partly dependent on the industry. The animals migrate in the summer to cooler pastures where there are fewer flies, and this means moving north as well as to higher altitudes, including Norway's coastal islands, although some spend the whole year in forests. Herders travel with them, and families have summer camps near the reindeer. The effect of snowmobiles is to make it quicker to reach stray animals, but petrol is costly and breakdown could lead to fatal exposure. The meat sells well in Scandinavia and Germany. At least one-fifth of the stock is culled each year, so with the average weight of a cleaned carcase at 40 kg, nearly 5,000 tonnes of meat can be produced annually. At a selling price of £1.20 or $2 per kg, a reasonable livelihood is made by an owner of 300 or more animals, but other occupations can be profitably continued. Governments have sought to protect the industry over the centuries, but vast hydroelectric schemes have flooded some of the best pasture lands. In Sweden, the Reindeer Breeding Act of 1971 sought to improve herding techniques and increase stock turnover by lowering the average age at slaughter from 8 or 9 to 3 or 4 years. Norway has established state-owned slaughterhouses.

Fisheries

The Atlantic coast of the northern cap has a settlement pattern which is primarily the result of a fish-based economy. Fishing has been, and remains, the mainstay of the Norwegian population. Almost all the towns are fishing ports in addition to their other functions, with Tromsø, Hammerfest, Svolvær, and Bodø as the main centres. There is a coastal fishery, within the fjords and inside the outer islands, an inshore fishery on the banks up to 100 km from the shore, and a distant water fishery extending out into the Norwegian and Barents Seas. In coastal waters and inshore cod is the main catch, with herring south of the Lofotens. Out to sea, cod, capelin, haddock, and pollock (saithe) are caught. Norway, which is the major fishing nation in all the North Atlantic waters, has of course an absolute predominance round north Norway. The three counties of Nord Norge account for nearly half the fish landings in Norway (1.39 million tonnes out of 2.9 million tonnes in 1972), and are the home of half the country's fishermen (17,800 out of 35,000 in 1971). This heavy reliance on fish leads both to a sensitiveness about the preservation of fishing rights, and also to an emphasis on the scientific management of the resource.

There is fishing also on the rivers and lakes of the interior. By comparison with the Atlantic fishery it is very small, but the catch, largely salmon and trout, is valuable. There is no significant commercial fishery in the Gulf of Bothnia.

Electric power

There are no fossil fuels yet exploited in the northern cap (but the chance of offshore oil is good). Power therefore means almost entirely hydroelectric power, and the major sources are the longer and slower-flowing rivers of the Baltic slope. The Kemi in Finland and the Lule in Sweden have been the sites of major developments. Each river has the greatest potential for hydroelectric power in its country, and each has a chain of stations put up in the years after the Second World War. The Lule system had in 1974 a capacity of 2,500 MW out of a potential of 3,300 MW, and the Kemi had in 1976 a capacity of 700 MW out of a potential of 1,000 MW.

There are some interesting bi-national schemes. The relatively small Pasvik river (Paatsjoki in Finnish), which forms the boundary between Norway and the USSR, has had virtually its full potential (100 MW) harnessed by joint action of Norway, the USSR, and Finland. Similarly a station on the Tuloma, in Soviet territory and feeding power to Murmansk, is a joint Finnish–Soviet project. There is a major interconnection between Letsi in Sweden and the Kemi system, with a transmission capacity of 600 MW; and two smaller ones between Tornehamn in Sweden and Sørnes in Norway, and Kalix in Sweden and another Kemi installation. There are other possibilities still unrealized, for both international and national power production. But the northern cap is already producing much more power than can be used locally, and the surplus is sent south. In Finland the question of whether to turn to nuclear power rather than to harness the remaining hydroelectric potential in Lappi is under official study.

Population

The northern cap has been populated since the retreat of the ice 10,000 years ago – probably a longer period than any other part of the northlands. Given its climatic advantages, this is not surprising. The people thought of today as being its aboriginal inhabitants, the Lapps, may be the descendants of the earliest dwellers in the region, or they may be later immigrants; the case cannot be proven either way. Although their language is Finno-Ugrian, and therefore originating somewhere to the east and south, that may have reached them when they were already in the north. In any case, the Lapps have been in the north a long time, of the order of several thousand years – but the ancestors of today's Norwegians and Swedes may have been there about as long. The northward movement of Finns took place much later, in the sixteenth and seventeenth centuries AD, and they pushed back the Lapps in front of them.

The earliest references to Lapps refer to them as Fenni, Finni, Scritofinni, and similar words. Finner is still the word for Lapps used in north Norway, where today's Finns are called Kväner. The Lapps' self-appellation is Sabme or Same, a name now gaining wider currency. But it is no insult to the people for English-speakers to continue using the English word Lapp, which has been in English usage, particularly in the longer form Laplander, since the sixteenth century.

The number of Lapps is not easily determined. As a people they are closely intermixed with their neighbours, and most of them are hard to distinguish by either personal appearance or occupation. There is much intermarriage, so that it is a problem to know who ought to be described as a Lapp. The generally accepted figures for 1975 were about 21,000 in Norway, 11,000 in Sweden, and 4,000 in Finland (there were also about 2,000 in the USSR). Thus the Lapps live under four flags – as do the Eskimos. The Lapps may be divided into groups defined by the way they live: sea Lapps, the largest group, fish on the Norwegian coast; mountain Lapps were the reindeer nomads who made long migrations and attracted the greatest outside interest; and the forest Lapps in Finland and the USSR have an intermediate economy with hunting, trapping, inland fishing, and farming as well as reindeer herding. The great majority of Lapps live in the northern cap, but some are to be found further south in the mountains of Norway and Sweden, even to the latitude of Trondheim. But the 36,000 are dispersed in a population, of the five northern-cap counties, of 923,000, so they constitute under 4 per cent of the total (see Appendices 1 and 2).

While the population is sparse by Scandinavian standards, at 2.5 persons per sq km, it is the densest in the northlands. There are a number of towns within a relatively small compass. The biggest is Luleå, the county town of Norrbotten (64,000), followed by Tromsø (43,000), Bodø (31,000), Kiruna (31,000), and Rovaniemi (28,000) (1974 figures). Of these five largest towns, three are coastal, reflecting the predominantly coastal economy, but two are well inland: Rovaniemi, the capital of Finnish Lapland, and Kiruna, an important mining centre. Tromsø and Luleå can claim to be cultural as well as economic centres, the first having acquired a university in the 1960s, the second a research institute. Settlement is discontinuous in the sense that there are significant empty areas (the *vidder*), but these are small by comparison with Siberia or Canada. The Norwegian coast is surprisingly thickly settled: a house is almost always in sight, a fact which will at once strike the traveller from inland.

There is much mobility in this population. There are marked seasonal migrations, attracted by fisheries, such as the cod harvest around Lofoten in January to March, or by lumbering in Norrbotten and Lappi. But many of the people who participate come from within the northern cap, having no work locally. Another, longer-term, movement has been across the frontier from Finland into Sweden, and this increased notably in the late 1960s. Finns believe there are better jobs and a better standard of living in Sweden. Some of this movement is brought about by a rationalization of forest working, permitting year-round operation and thus reducing seasonal labour. But the most serious trend, as far as the governments are concerned, is the movement right out of the north – the retreat from the frontier. It was noted in Norrbotten in the 1950s, and then in Norway and Finland too. In Norway and Finland the effects of the devastation created by the German scorched-earth policy in 1944–5 were far-reaching. It is expected that the total population will not grow over the next decade or so; at best the high birth-rate (now diminishing) may prevent a drop. The southwards movement is not always just the familiar drift to the towns and abandonment of rural life, because many emigrants come from the larger centres. While the cause may be closely linked to the developing economies of the three countries, it is also true that some people have simply had enough of the winter cold and darkness.

If we think of the problem of providing a labour force, we must bear in mind the fact that the northern cap, unlike almost all other parts of the northlands, has a local population which has been settled there for many years and is skilled in the ways necessary for the traditional economy of the area. The problem with labour, therefore, is more how to retain it than how to attract it. This had led to emphasis in all three countries on ways of making life more attractive, with special programs for education and social services.

Some specialists have to be brought in, however, because many skills are lacking in the northern population. It is the unskilled labour which is sufficiently abundant. Training and retraining courses are organized locally, especially in Sweden. The university of Tromsø was established for this reason, and one of its main purposes is to provide a northern centre for medical training.

Government and administration

The three countries – Norway, Sweden, and Finland – are all independent sovereign powers, and are all social democracies in the Western tradition. But within this framework there is much diversity. Culturally, the Norwegian and Swedish languages are mutually comprehensible, but Finnish is totally different. Historically, Norway and Sweden have been closely interlinked – they were a joint kingdom between 1814 and 1905 – while Finland was a Grand Duchy of the Russian Empire from 1809 to 1917. In the political context of today, Norway is a member of NATO, Sweden is non-aligned, and Finland has a special relationship with the USSR. These factors pull in different directions, but there is also increasing integration between the three countries. The Nordic Council (Nordiska rådet), founded in 1952, brings together government representatives of all the Scandinavian countries. It has assisted in the creation of a common passport area, a common labour market, and integration of national health schemes. It has also promoted both a Nordic Lapp Council (Nordiska samerådet, 1956), which organizes a Nordic Lapp conference every three years, and a northern-cap conference which has met regularly since 1962. Proposals have been made for strengthening the international ties in the north; in particular, the possibility is being examined of creating some kind of federal administrative district in areas where the frontier divides an economic unit – for instance, the lower Torne valley, where the river is the frontier.

The various frontiers which criss-cross the northern cap are not of great antiquity. The Norwegian–Swedish frontier, which runs chiefly along the height of land, was determined in 1752–6; the Swedish–Finnish frontier along the Muonio and Torne rivers in 1809; the short section of Norwegian–Soviet frontier on the Pasvik river in 1826; and the Finnish–Soviet frontier to the south of that in 1944. This last reflects the most recent redrawing of the international boundaries in the area, at the end of the Second World War, when Finland lost the outlet to the Arctic Ocean (the 'Petsamo corridor ') she had had since 1920. The Soviet frontier with Norway and Finland, which constitutes the eastern boundary of the territory we are considering here, is quite different in kind from the others. It has barbed wire, a wide prohibited zone, very few crossing points, and those very seldom used; while the others are generally unnoticeable on the ground and are crossed daily by as many people as wish, generally without formality or documents of any kind. But even the latter sort of frontier does, of course, form a legal divide. Before the creation of the common passport area there were special agreements covering the movement of reindeer and herders.

Local government in the three countries is broadly similar. Elected county and parish or commune councils administer local affairs. There are national differences in the powers and administrative style of the councils, but these are not very significant. One real difficulty arises from sparseness of population, which tends towards administrative units which are large in size and low in effectiveness. Population centres often have too few inhabitants to support adequate social services. A consequence of the tri-national division is that there is no one dominant town which might act as a capital and provide big-city services.

A rather striking feature of the northern cap has been its relatively high communist vote. In general elections in Sweden in 1968, 1970, and 1973 the communist vote in Norrbotten was nearly 14 per cent – the highest for any county and about three times the national average. It had been higher in earlier elections. In the 1972 and 1975 elections in Finland, Lappi likewise had the highest communist vote – about 30 per cent, while the national communist vote was 18 per cent. A comparable situation existed in north Norway in the 1950s and early 1960s, but there the communists split into two factions, lost many votes, and in 1973 did not contest the parliamentary elections at all. The reason for the high communist vote in the north perhaps lies more in economic stagnation, of which remoteness and isolation are part causes, than in sympathy with, or proximity to, the USSR.

Finally, a major binding factor is the presence of the Lapps, a national group with its own distinctive culture and language.

Transport

The northern cap, like other parts of the northlands, is a region first served by water transport. Norway operates coastal steamers, a service which long provided the only link between the towns and villages of her long coastline, and it still plays a basic role. Every village is a port, but the main centres at which the express steamers call daily are Brønnøysund, Bodø, Svolvær, Stokmarknes, Harstad, Finnsnes, Tromsø, Hammerfest, Honningsvåg, Vardø, and Kirkenes. On the Baltic side there are ports at Piteå, Luleå, Kalix, and Kemi, but there is no regular passenger service to them, owing to both the presence of ice in the winter and the easier conditions for overland transport. The Baltic gives access to an important system of rivers – Pite, Lule, Torne, Kemi – which stretch far into the interior and may be used for shipping and timber-floating; but river use has declined in recent years as hydroelectric schemes have been introduced.

Railways in the northern cap are rather few. Much of it is unsuitable railway country. The first to be built was a line from Luleå to the mines at Gällivare, completed in 1887. It was joined to the Swedish national network a few years later, and then extended to Kiruna and across the mountains into Norway at Narvik in 1903. This was and is the most important line in the region, for it carries the iron ore from one of the world's major sources to an ice-free Atlantic port, and in fact carries the heaviest traffic of any line in Scandinavia. Meanwhile, the Finnish railway network was extended in 1903 to Tornio, and this was at the Russian broad gauge (still standard for all Finnish lines). The only additions built since then are a continuation of the Swedish coastal line to the Finnish frontier (1913), and further extension of the Finnish network up the Torne to Kolari and up the Kemi to Rovaniemi and beyond to the Russian frontier. The north Finnish system carries chiefly timber. Thus the northern half of the cap area has no rail system at all.

Roads are better developed, but the improvement is recent. Military operations in the Second World War stimulated much construction. The main features of the road network are these. The Norwegian coastal road, which continues from the south right up to the Soviet border at Jakobselv, is not only the main artery of the country, but, for much of its length in the north, the only artery, there being no alternative land route. Several arms of the sea have to be crossed, and for this ferry boats are used. The Norwegian and Swedish systems are not connected by major direct links. There is a minor road from Mo i Rana to Umeå, while the more important link, to the north, crosses from Skibotn into Finnish territory, whence several roads cross the Muonio river into Sweden. The Finnish north has quite a good network, and includes four other links with Norway (Enontekiö–Kautokeino, Inari–Karasjok, Inari–Tana, Inari–Kirkenes). The main 'Arctic highway' in Finland, running up through Rovaniemi and Ivalo, is a well-maintained year-round road. In Sweden the main artery runs along the Baltic coast, where there are no mountains to complicate construction, and the area inland is well served until the mountains are reached, when the roads almost disappear. Once again, if the comparison is between the northern cap and the rest of Scandinavia the northern cap is much more poorly served, but if it is with the rest of the world's northlands, then the cap is rather well off.

Almost contemporaneous with the road improvements is the air service. It expanded rapidly after the Second World War, so that today virtually all centres with over 10,000 inhabitants, and many with less, have their own airfield and regular flights. The service in each of the three countries developed independently, as might be expected, so there has been little integration. Thus until 1975 both Norwegian and Finnish airlines flew to Kirkenes, and the quickest route from Kirkenes to Oslo was via Helsinki; but the Finnish link has now been stopped. There is nothing in the northern cap to compare with the widespread use of private light aircraft found in Alaska.

Neither by water, land, nor air is there any significant link between the northern cap and the USSR. The road system crosses the frontier just outside Kirkenes, and farther south, at Raja-Jooseppi between Ivalo and Murmansk. Some of these roads are relics of the Finnish corridor to the Arctic Ocean. But today they are normally shut. The Finnish railway line to the frontier near Salla does not run through, as originally intended, to join the line to Murmansk; it has never been completed. There is no commercial air link north of that between Helsinki and Leningrad. The only encouraging signs in the direction of opening up this frontier have been the permitted passage of some tourist parties crossing the frontier at Kirkenes and Raja-Jooseppi; but at the latter only Finnish citizens may pass through.

Significance of the northern cap to Scandinavia

Each of the three countries sees its northland as a kind of depressed area, but each sees it in a different light. Many problems are common to all three sectors: remoteness, sparseness of population, hard climate. But some other problems, and most of the advantages, are peculiar to each.

For Norway, the north is more important than it is for the other two. Not only is Nord Norge a bigger proportion of the country in terms of area (35 per cent, compared to Lappi, which is 30 per cent of Finland, and Norrbotten, which is 22 per cent of Sweden), but the Atlantic seaboard gives Norway wider possibilities of overseas contacts, has important fisheries, and good prospects for oil. The location of Nord Norge likewise gives it exceptional strategic significance. In the Second World War this fact brought about the devastation of much of the area by the German army in 1944–5, and thus also the need for costly reconstruction. Today, the strategic significance of the area is a prime reason for Norway's membership of NATO, and there is considerable defence investment.

While Finland also suffered extensive devastation in the north in the war, and emerged from it further handicapped by the loss of territory, her remaining northern territory is much less important strategically than Norway's, having no ocean outlet, and there is no defence expenditure there. But 'the north' has more emotional overtones for a Finn than for a Norwegian or a Swede, in that there are many Finnish-speaking districts in the northern cap outside Finland itself, and many places in north Norway and Sweden have Finnish place-names as well as their own (thus Jälivaara for Gällivare, Vesisaari for Vadsø). Despite Lappi's lack of resources, its remoteness, and poverty, one can imagine no circumstances in which Finland might agree to part with it.

From the point of view of resources, Sweden is at present the luckiest of the three northern-cap powers. The rich iron ore is at the base of Norrbotten's relative prosperity, and as a result there is a better infrastructure and a more varied spread of local industry. Sweden's neutrality in the war also gave her a significant advantage, in economic terms, over her two partners. Nevertheless, there is a southwards drift of population. There are too many people for all to have jobs in the area, but too few to provide a large enough base for services.

So, although the northern cap is a problem area, causing concern in Oslo, Stockholm, and Helsinki, its problems are being tackled and there are real possibilities for a positive outcome. Isolation is declining and political and economic co-operation is increasing. There is the probability of continuing rural depopulation, but it is being squarely faced, and strategies of retreat are being worked out. If military tension can be reduced, links with the USSR may grow; there could be expanded coastwise trading with Norway, timber exports to Finland, and no doubt much else. Tourism, already considerable, could be increased. One must remember that a problem area in sophisticated and technologically advanced Scandinavia is not so severe when seen in a wider context. Indeed, the northern cap certainly has fewer difficulties than the other northern territories we are considering, and there are undoubtedly ways in which Scandinavian experience can be helpfully applied in similar situations elsewhere.

Svalbard

Svalbard (meaning 'cold coast') is the name given to the group of islands some 700 km north of Norway, situated on the continental shelf and therefore to be regarded as an extension of Scandinavia. The archipelago comprises Spitsbergen (formerly called Vestspitsbergen), Nordaustlandet, Edgeøya, Barentsøya, Prins Karls Forland, Hopen, Kvitøya, Bjørnøya, and other small islands. Norway exercises sovereignty, conferred on her by a treaty signed in Paris in 1920, and is obliged by that treaty to accord to any other signatories the same rights as herself in the exploitation of natural resources. The treaty also requires that the region may not be militarized. At present, the only country besides Norway to exercise its right of mineral exploitation is the USSR.

The rocks of the islands are strongly folded in many regions, and the highest mountains are Newtontoppen and Perriertoppen in northeast Spitsbergen (1,717 m). There is much contemporary ice cover, which covers about 50 per cent of Spitsbergen and 75 per cent of Nordaustlandet, but it was even more extensive in the Pleistocene. Many glaciers reach the sea. The landforms are largely Alpine and the coast fjorded. A low platform, or strandflat, stretches along part of the coast, and is in some places more than 10 km wide.

There never have been any indigenous inhabitants of Svalbard. It was first brought to public attention by Barents in 1596, but was almost certainly known to Norse and Russian seamen before that. It was initially confused with Greenland, and called by that name. Whaling was carried out in Svalbard waters in the early seventeenth century by the Dutch and the English, and after that trappers, first Russian and then Norwegian, were active on shore. Whalers even set up the town of Smeerenburg in the seventeenth century. Coal was discovered at this time also, but there was no mining until 1904. Permanent settlement began only with the initiation of mining operations. The population in recent years (1970–4) has been between 2,800 and 4,200, made up almost entirely of people employed in the coal-mines. Of this total, the number of Norwegians remained stable at about 1,000, while the number of Russians varied between 1,800 and 3,200.

The coal has been much the most important economic factor in Svalbard's history since Norway acquired sovereignty. It has special importance for Norway herself, because the Norwegian mainland is totally lacking in coal. A Norwegian company now operates mines at Longyearbyen (named after one of the original American operators, Mr Alfred Longyear) and Sveagruva. Another Norwegian company had a mine at Ny-Ålesund, but it was closed in 1962. Other countries have had claims, but only the USSR has been active since before the Second World War. The Soviet workings are at Barentsburg, Grøndal and Pyramiden. Grøndal adjoins Barentsburg, and the Russians operate there under a lease from the Norwegian company signed in 1971. All these localities are in the Isfjorden area of Spitsbergen except Sveagruva, which is on Van Mijenfjorden, the next fjord to the south. Shipments of coal totalled between 800,000 and 900,000 tonnes a year in 1969–74, with the Norwegian and Russian mines accounting for about half each, despite the larger Russian population. While these figures show little growth, it is quite possible there

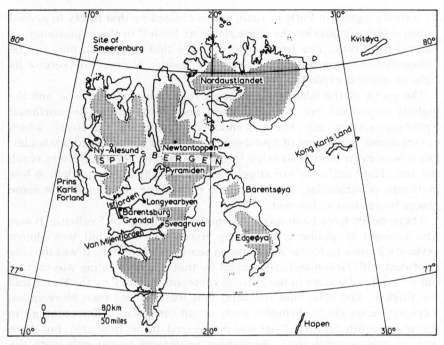

Map 18. Svalbard. Shading indicates ice caps

will be growth in the near future, for both the Russian mine at Grøndal and the Norwegian mine at Sveagruva are in the early stages of development.

Shipment is of course subject to the limitations imposed by floating ice. Until recently this has meant a shipping season of about six months (June–December), but more powerful and better designed ships have extended it further. A Soviet ship reached Barentsburg on 20 January 1976.

Although other minerals have been found in Svalbard – gypsum, lead-zinc, marble – and some attempts have been made to exploit them, none of these deposits has been worked for many years, and there is little prospect of development now. But there is one most important mineral prospect: oil. Exploration started in the late 1960s, and continues. Interest now centres on offshore areas, but the island of Hopen has also been the site of drilling. No oil or gas has yet been found in commercial quantities, but the area remains promising.

Biological resources were exploited quite vigorously in the past: first whales, later Arctic fox, polar bear, reindeer, seal, and walrus. But apart from the whaling and sealing, this was never an organized industry, but rather the occupation of a few tens of individuals working for themselves or in small groups. Uncontrolled hunting (it must be remembered that until 1920 Svalbard was *terra nullius*, belonging to no one) led to severe reduction of certain populations, such as the reindeer, which were nearly exterminated by the end of the nineteenth century. This gave rise to protective legislation for them, and later also for walruses and polar bears. Muskox were introduced in 1929 but are not hunted. Today only the small fur-bearers are trapped, and the scale of operations is small. The important biological resource of the area is the fish of the surrounding waters. Nearly 600,000 tonnes were taken there in 1974 (see Chapter 7).

Until recently the only way of getting to Svalbard was by sea, and thus subject to the ice limitations already mentioned. For some years after the Second World War there were irregular flights to a makeshift airstrip at Longyearbyen, usable only in winter, but in 1975 an airfield was built, and there are now regular flights from Norway (twice a week) and the USSR (once or twice a month). The reason why the link was not established much earlier was that Norway was unable to secure Soviet agreement that provision of an airfield did not constitute militarization forbidden under the treaty of 1920. The relative proximity of the USSR's major naval base at Murmansk increased the sensitivity of the issue.

The establishment of regular flights has stimulated the tourist industry. The potentialities of tourism in this highly scenic land have long been recognized, and a hotel was run for some years from 1896. Cruise ships included Svalbard on their circuit, and from 1935 there has been a regular ship service from Norway. The time-saving and the avoidance of rough seas offered by air transport have increased the traffic.

Quite apart from the economic activity, and exceeding it in importance, is the strategic issue. The seas between Svalbard and Norway must be traversed by almost every unit of the Soviet northern fleet based at Murmansk (very few head eastwards along the northern sea route), as well as by the very large fishing fleet also based there. Svalbard's importance in this context became apparent in the Second World War, when Allied convoys were carrying large amounts of freight to Murmansk. A small Allied force operated in Spitsbergen in 1941 and 1942–5, to prevent the establishment of a German base. This object was achieved, but German naval bombardment destroyed existing facilities in the mining settlements, and clandestine weather stations were manned for much of the time.

Currently, the most important argument concerns the division of the Barents Sea continental shelf between Norway and the USSR. If Svalbard is regarded as Norwegian territory, the median line (joining points equidistant from the nearest land of each power) would lie somewhat to the east of the meridian of Vardø. But the Soviet argument adduces various ' special circumstances' and sticks to the sector boundary shown on Soviet maps since 1926, allowing a larger slice to the USSR. Significant factors

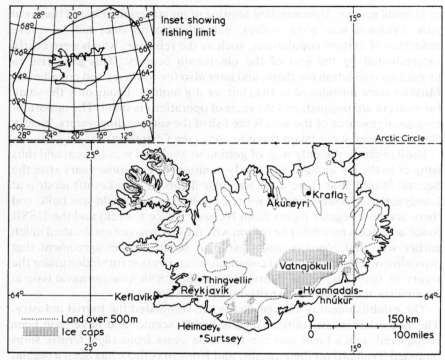

Map 19. Iceland

in this argument are the islands of Bjørnøya, 220 km to the south of Spitsbergen, and Hopen, 100 km southeast of Edgeøya, both of which form part of the Svalbard group.

Bjørnøya, a desolate island about 20 km by 15 km, discovered by Barents in 1596, is important now chiefly for its weather station, but in 1915–25 a small deposit of coal was worked there. It may, like Hopen, have a future in the oil industry.

Iceland

Iceland is included in the territories covered by this book more on account of its relations with other northern lands than for its own attributes, for it is the least 'Arctic' of the countries we are considering. It is the only one of the eight circumpolar countries which lies wholly south of the Arctic Circle, and it is anomalously warm due to the influence of the North Atlantic drift. Iceland's climate has been characterized as cold temperate oceanic, with a high positive anomaly in the winter half of the year. On the other hand, although the country is farther from the North Pole than any of the others, its southern limit is also farther north, and the fact that it thus has no southern base to operate from gives it a distinctive northern flavour.

The country lies between latitudes 63° and 66° N, and measures about 350 km from north to south and 500 km from east to west. Its rocky surface is young, no part of it more than 20 million years old and 10 per cent of its lavas less than 10,000 years old. The ridge running down the middle of the Atlantic crosses Iceland, which is the site of many volcanoes. Some are still active, and there have been two recent dramatic volcanic events off the south coast of Iceland: the appearance of a new island, Surtsey, in 1963, and an eruption on Heimaey in 1973. The highest elevation in Hvannadalshnúkur in southeastern Iceland (2,119 m), a dome in the largest ice cap, Vatnajökull, which has an area of 8,400 sq km and a maximum thickness of 1,000 m. Permanent ice covers 11.5 per cent of the country. The landscape is almost treeless, probably due more to human action than to natural causes. Iceland is indeed 'the land of ice and fire', and is not, consequently, an easy place for humans to support themselves.

Iceland was probably seen by a Greek inhabitant of Marseilles, Pytheas, who sailed northwards about 320 BC. But, as with some later explorers, his account was dismissed as fanciful and it had little impact. Iceland was colonized from Norway in the ninth century, and has a recorded history from that time. The first colonists found Irish hermits already there, but before them (and they cannot have been there very long) there was no human habitation. The country was initially a republic, but from 1262 it was associated with Norway, and later with Denmark. Although Iceland always had considerable freedom in domestic affairs, formal independence was obtained only in 1944.

Resources

Because of its geological history, Iceland is very poor in exploitable mineral resources. In the past sulphur, Iceland spar, and bog iron were exploited, but none are worked now. Today the main extractive industries are for volcanic glass and pumice, used in the building industry because of their insulating properties.

The absence of fossil fuel is compensated for by other forms of energy. Iceland is exceptionally rich in natural heat sources. There are hot springs, steam-fields, and large areas of heated rock, and all of these can be used in various ways. The hot water has been used for washing and bathing since earliest times. Today piped hot water is organized at municipal level for Reykjavík and many other communities. The heat source is now being used for the generation of electricity, at a plant at Krafla in the northeast. There is considerable hydroelectric potential in the rivers. About 5 per cent of this is at present harnessed, but there is rapid growth (output increased by 70 per cent between 1964 and 1975). There has also been a major and successful effort to bring electricity into rural areas. Besides these two energy sources, there are large peat deposits, which have been much used in the past for domestic heating, but are not the object of large-scale exploitation now. There is also some possibility of offshore oil, particularly off the northeast coast.

The renewable resources of Iceland are the country's chief asset, and the most important of these is fish. The waters round Iceland are especially rich in fish, for it is an area where warm and cold currents mix. Cod and herring became the main catch, but not to the exclusion of other species, such as haddock and redfish. The herring catch has dropped greatly in the last few years, and has been replaced by capelin, which even overtook cod to reach the top of the catch table in 1973. Other nations fish in the vicinity of Iceland, as the 'cod wars' have demonstrated. Britain and West Germany are the two biggest after Iceland, but Iceland's share of the catch has steadily increased from 45 per cent in 1953 to 81 per cent in 1975. The absolute figures for 1975 were Iceland 907,000 tonnes, Britain 116,000, West Germany 65,000, out of a total of 1.11 million tonnes caught in the area designated 'Iceland waters' (Va) by the International Council for the Exploration of the Sea (see Table 33). Much of the fish caught by Iceland is exported to a market built up over many centuries, and it constitutes not only the major but almost the only item in the country's exports. The total Icelandic catch in 1975 was 995,000 tonnes (the difference being made up outside Iceland waters), which puts Iceland between Canada and France in the international table.

The reasons for the conflicts over offshore fishing have been basically the growing pressure on fish stocks, and the tendency for commercial fishing interests of several nations to invade spawning grounds (of which there are many round Iceland) when unable to meet their catch requirements elsewhere. Iceland has sought to give the stocks more protection than her partners were willing to concede. To do so she proclaimed in 1975 an exclusive economic zone of 200 nautical miles – a greater extent than was then permitted by international law. But since her economy is based so heavily on fish, there was much sympathy for her cause, and in any case the idea of a 200-mile exclusive economic zone soon gained general acceptance.

The country's other renewable resource is agricultural products. Although the terrain is scarcely promising, the soil is generally fertile. There is very little arable farming, but grasslands are now cultivated and holdings of sheep are considerable (865,000 in 1974); cattle and horses are also kept, but on a much smaller scale (67,000 and 44,000 in 1974). Dairy farming is important, however, and milk yield is high at about 3,000 litres per cow per year. Horses are to some extent a survival of the past, when the tough Icelandic pony was the only means of transport. There is some market gardening, but even the potato crop is not sufficient to meet home demand. There is a growing industry in greenhouse cultivation, the heating being provided from natural sources.

Population

The population of Iceland in 1975 was 219,000. This is the largest it has ever been, and it is increasing at the rate of about 1 per cent a year. But the growth has been steady only since 1800 (apart from loss by emigration in the late nineteenth century). It is believed that the population was about 75,000 in 1100 and the same in 1300 and 1900, while in 1800 it was 47,000. The small isolated community was vulnerable to infectious disease and famine. The original stock of Norsemen with some Celts has been little augmented by recent immigration, while there has been some emigration, chiefly to Canada, the USA, and Denmark.

The pattern of settlement is round the coasts, with the centre of the island, which is largely a lava or ice desert, virtually uninhabited. Over half the population lives within 15 km of Reykjavík, and the other towns are all under 12,000 inhabitants, Akureyri on the north coast being the chief among them. Most of the settlement locations are of long standing.

The population is socially homogeneous and there is almost universal literacy. Culture and scholarship have always been highly regarded, as one might expect in the land of the sagas. Very special efforts have been made to preserve the purity of the Icelandic language, and loanwords for such concepts as telephone or television have been rejected in favour of words from Icelandic roots.

The manpower problem in Iceland is different in kind from that elsewhere in the north. Icelanders have very high educational and vocational standards, and there is no pool of untrained 'native' labour. Icelanders have learnt to be flexible in their work pattern, and many equip themselves with quite diverse skills; thus an expert in avalanche control is also a skilled plasterer. Since the island is the whole nation, and there is no 'south' to call on, many of the jobs have no counterpart in other northern territories. In 1973 employment was distributed by cost as follows: agriculture 10.7 per cent, fishing and fish industries 12.8 per cent, other manufacturing 17.5 per cent, commerce 13.9 per cent, services 23.2 per cent, construction 12 per cent, transport 8.6 per cent.

It is not possible to analyse the significance of the north in Iceland's economy; the economy is in our sense wholly northern. One can only say that that economy, managed by a stable government, has produced a high standard of living for the people.

Government and administration

Despite the poorness of natural endowment, the community of Icelanders has a long tradition of stable government. The country claims the earliest parliament, the Althing, created in 930, which met in the open air at Thingvellir, near modern Reykjavík, for many hundreds of years. Today its sixty members meet in Reykjavík and carry forward a parliamentary democracy organized on party political lines under a president.

Local government is exercised through a system of sixteen counties (sýsla), each headed by a sheriff (syslumaður). The nineteen towns have their own councils and are independent of the counties. The earlier subdivision into four provinces (fjorðungur) – northern, southern, eastern, western – no longer has any administrative function.

At the international level in Iceland is a member of the United Nations, a member of NATO, and trades with many countries of the world. Membership of NATO has been an important factor, for it aligns Iceland politically with the Western world. However, its relevance to the country is now being called into question by some Icelanders.

Transport

Sea transport is naturally basic to Iceland. Since the country is by no means self-sufficient, the necessary imports, and exports to pay for them, must move by sea. The network of services is effective and long-established. Most years the ice edge remains north of Iceland and then all ports are ice-free the year round, but sometimes it advances to the coast and so blocks coastal traffic. Such obstruction was frequent before 1918, and then, after a relatively open period, started to become frequent again in 1965.

Land transport has always been difficult in this terrain. Until less than a century ago pack-horse paths were the only roads, and there were virtually no wheeled vehicles. Now one in three Icelanders has a car, a road network includes all the towns, and it is possible to drive the whole way round the island (the last section of 33 km across the outwash plain of Vatnajökull, an area subject to sudden flooding, was completed in 1974). Inevitably the roads are narrow, and the surface is gravel or volcanic slag; but expenditure on roads and bridges is higher per head of the population than in most countries. There are no railways.

The difficulties caused by sea ice and by a restricted road system were of course mitigated by the arrival of air transport in the 1920s. There are now scheduled domestic services from Reykjavík to all the major towns, in some cases several a day. They are operated by Icelandair (Flugfélag Islands), which also has services to Western Europe and, in summer, Greenland. An associated company, Icelandic Airlines (Loftleiðir) runs longer-distance flights, particularly to New York. By not being a member of the International Air Transport Association it is able to undercut other transatlantic airlines and offer a stop-over in Iceland as well.

The importance of the air link in military matters is demonstrated by the NATO airbase at Keflavík. Iceland lies in the middle of that arm of the north Atlantic which must be traversed by Soviet naval units leaving Murmansk. The island's earlier importance as a stepping-stone across the ocean has dwindled as aircraft range has increased.

Jan Mayen

Jan Mayen is an elongated island some 55 km in length situated 550 km north of Iceland. It consists chiefly of the extinct volcano Beerenberg, a glacier-covered cone rising to 2,277 m. It is named after the Dutchman Jan May who saw it in 1614, when Dutch whalers were often in the vicinity, but others very likely saw it earlier. There are no native inhabitants, but hunters and scientific expeditions have visited the island from time to time. Since 1921 there has been a Norwegian weather station there, and Norway obtained sovereignty over it in 1929.

The Faeroes

The group of rocky islands known in English as the Faeroes, in Danish as Færøerne, and in Faeroese as Føroyar, lie roughly midway between the north coast of Scotland and Iceland. They occupy an area about 120 km from north to south and 70 km from east to west. The climate is oceanic, without much difference between seasons, with heavy precipitation and frequent storms. The highest point in the islands is Slættaretindur (882 m), and there are spectacular cliffs.

While there are some indications of Celtic settlement as early as 600 AD, the islands were colonized about 800 by Vikings from Norway, and remained part of Norway until the seventeenth century. A connection with Denmark then grew, and remained when the union of Denmark and Norway was dissolved in 1814. In 1948 the islands became largely autonomous, but continued to acknowledge the Danish crown. The population,

40,000 in 1974, speak Faeroese, a language related to Norwegian and Icelandic. The treeless landscape contains grasslands on which sheep are kept. There are also some cows, but very little arable farming. Much the most important activity is fishing, and Faeroese fishing boats operate in Iceland and Greenland waters as well as offshore locally. There is also some shore-based whaling, the pilot whale frequenting Faeroese waters in summer. The capital is Thorshavn.

CHAPTER SEVEN

The circumpolar oceans: international conflict and co-operation

The North Atlantic

The part of the North Atlantic with which we are concerned lies north of a line running from the southeast corner of Labrador to the southern tip of Greenland, Iceland, and the most westerly point of Norway. It therefore comprises the Labrador Sea, Baffin Bay, Denmark Strait, the Greenland Sea, the Norwegian Sea (officially the Greenland Sea is part of the Norwegian Sea), the Barents Sea, and the White Sea. The waters to the north of these seas are included in the section on the Arctic Ocean.

This area is at the southern edge of the main Arctic pack-ice zone; about half of it is ice-covered for part of the year (see Map 4), but only the northwest corner of the Greenland Sea has permanent ice cover. The permanently open half includes the north coast of Scandinavia, and part of the southwest coast of Greenland. This is important for shipping. Also important for the same reason is the fact that the pack-ice belt which closes the Labrador coast and the St Lawrence estuary is present for a relatively shorter time than most of the rest of the pack ice – about three months, January, February, and March.

The northwest Atlantic is the major iceberg area in the northern hemisphere. The icebergs originate chiefly on the west coast of Greenland, whence they drift across Baffin Bay and down the Labrador coast, as the current carries them southwards. North of Newfoundland they fan out into the Atlantic, drifting southwards and eastwards until they melt away. Thus icebergs may be encountered anywhere in a rather large water area, between May and September. Because they may drift across some of the principal shipping routes, they constitute a major hazard. Everyone remembers the sinking of the *Titanic*.

The distribution and movement of pack ice and particularly of icebergs are controlled mainly by currents. There are two major currents in this region, one warm and one cold. The warm current is the North Atlantic drift, which flows northeastwards across the North Atlantic. It continues round the north of Scandinavia into the Barents Sea, where it drops below the surface but flows on into the Arctic Ocean to become the layer of Atlantic water detectable at many points at a depth of 250–900 m. The cold current is the East Greenland current. It is the main outflow from the Arctic Ocean, and it loops its way round the tip of Greenland, up the west coast, where it is known as the West Greenland current and is warmer, round the head of Baffin Bay, and down the coast of Baffin Island and Labrador. This is the current, of course, which bears the icebergs and the pack ice into the northwest Atlantic. Both currents are relatively constant the year round.

Shipping and air routes

The main transatlantic shipping arteries skirt the southern edge of the area we are considering; but it contains a number of shipping lanes of lesser importance. The chief of these is the route round Scandinavia into the Barents and White Seas. It is used by the very considerable traffic to and from Murmansk, which is the USSR's major ice-free port. It is also used by traffic bound for the White Sea, Arkhangel'sk in particular, which is open for about six months in the year, and by the small number of ships which ply between the northern sea route and non-Soviet ports. The seaway leading eastwards out of Arkhangel'sk and Murmansk, across the southeast corner of the Barents Sea, is used quite extensively in summer by Soviet shipping bound for the Pechora and for the northern sea route.

Few freighters penetrate farther north in the Barents Sea. Occasionally the more northerly entrances to the Kara Sea may be used – Matochkin Shar, the narrow strait dividing the double island of Novaya Zemlya, and round the northern extremity of the island – but this only occurs when ice blocks the southern entrances. Supply vessels visit stations on Novaya Zemlya itself, including particularly the no doubt rather large station where Soviet atomic weapons have been tested, and also on Zemlya Frantsa-Iosifa. The main freighter traffic in the northern Barents Sea is that to Svalbard. The two mining communities there, Norwegian and Soviet, have a sea link with their parent countries for at least six months in the year (the proximity of the North Atlantic drift gives the west coast of Svalbard a longer ice-free season than the White Sea has). Passenger ships carrying summer tourists to see the midnight sun go to Svalbard and Zemlya Frantsa-Iosifa; Western ships to the former, Soviet ships to the latter.

The Norwegian coast is a busy area for shipping. Most of it is coastwise traffic, but Narvik plays an important role as the main outlet for the iron ore from Kiruna in Arctic Sweden, and Kirkenes has its own source of iron ore in the mine just outside the town. Both these ports therefore have a significant export trade.

Iceland need not concern us, since all its main sea links lie outside our area.

Greenland generates some coastwise traffic in summer, but the important link is with Denmark. The normal route is to Godthaab, and since this lies on that part of the southwest coast which is never closed off by ice, there is the possibility of year-round movement of traffic. In fact, winter voyages were not made until 1959, and then, most unfortunately, the new ship chosen to open the winter service, *Hans Hedtoft*, foundered on her maiden voyage after apparently hitting an isolated block of ice off the southern tip of Greenland.

Somewhat analogous to the Greenland coastal traffic is that on the Labrador coast. Here, too, there are a number of coastal villages with no road link between them and so requiring occasional visits by ships.

There remains the more frequently used, but seasonal, route which traverses the Labrador Sea and leads either to Hudson Strait, for Hudson Bay or Foxe Basin, or to Baffin Bay for points farther north. All traffic bound for the eastern end of the northwest passage must pass this way: for Frobisher, Resolute, Eureka, Igloolik. But the heaviest traffic is that for Churchill, on Hudson Bay, from which grain is exported to Western Europe. This trade has been continuing since 1931, and in 1974 thirty ships carried 22 million bushels (600,000 tonnes) of grain away from Churchill. A high point had been reached in 1971, with thirty-six ships carrying 26 million bushels. The season is from late July to mid-October.

There are various air routes across this territory, mostly of minor or local importance. The heavy traffic of the northern transatlantic routes just touches the southern edge (as with shipping) in the vicinity of Greenland. Flights from Europe to North American west-coast destinations (between Vancouver and Los Angeles) follow a more northerly course and cross Denmark Strait and Davis Strait. In the 1950s and 1960s there were intermediate stopping points at Søndre Strømfjord in West Greenland and Frobisher in Baffin Island, but these became unnecessary with increased aircraft range. The role of the north is restricted to emergency support.

Fisheries

Whereas these northern seas are traversed by relatively few regular freighting routes, they are extensively visited by fishing vessels. The Barents Sea, Norwegian offshore waters, and Iceland waters are fishing grounds of major importance, and the Labrador Sea and Svalbard waters also yield significant catches. The northwest Atlantic and the northeast Atlantic fishing areas taken together (i.e. the whole Atlantic north of latitude 35°) account for about a quarter of the world catch of fish at sea. But these areas include more of the North Atlantic than we are concerned with. Table 33 shows the catches in an area which coincides, as closely as published statistics allow, with our area, and which produced about a tenth of world sea catch in the 1950s and 1960s (Map 20). Its importance relative to the rest of the world has dropped in the last decade, although the amount of fish taken has increased. The leading country in the fishery is Norway, with the USSR next in importance, and Iceland third (see Appendix 5). The USSR has become a major fishing country only since the Second World War, but had fished the Barents Sea for some time before that. This order of priority has therefore probably been true for several decades, and is not likely to change soon. Of the other countries participating, Portugal is perhaps the most unexpected. However, her northern fishery, which is in the Labrador Sea region, is diminishing.

Of the species caught, cod long predominated, but its relative importance fell from over half the total catch in the area in 1953 to about a quarter in 1975, the cod catch being surpassed by that of capelin, a species not taken in significant quantities in 1962 or 1953, but now apparently replacing herring. In 1971 there was a relatively large catch (348,000 tonnes) of polar cod in the Barents Sea. Almost all this was taken by the USSR, probably for fish meal. The fishery has continued, but at less than half that level.

Sea mammals are taken in the northern North Atlantic. The principal industry is of the harp seal (*Pagophilus groenlandicus*) which exists in three populations, all within our area: the Labrador coast, Jan Mayen, and the White Sea. These are harvested by Canada, Norway, and the USSR. The total number taken averaged 200,000–300,000 animals a year from 1965 to 1974. The whaling industry is on a smaller scale. Until 1973 whaling was conducted from shore stations in Norway, Labrador, and Iceland. Table 34 shows that whale catches have been dropping since 1970, and the number of shore stations has correspondingly decreased from eleven in 1969 to four in 1973, all in Iceland.

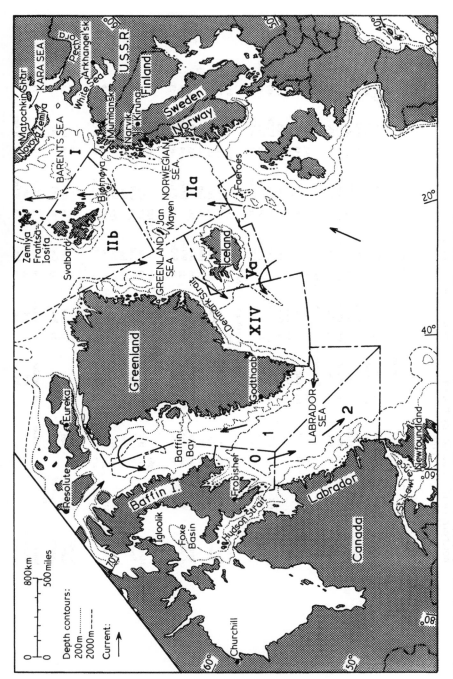

Map 20. North Atlantic fishing grounds. The numbers refer to fishing areas listed in Table 33

Table 33. Commercial fisheries nominal catch (live weight equivalent of fish landed) in northern North Atlantic, by fishing area, main species and main countries, 1953, 1962, 1971–75 (thousand tonnes)

Barents Sea (ICES sub-area I)

	1953	1962	1971	1972	1973	1974	1975
TOTAL	279	1,996	823	1,155	1,110	2,056	1,817
Main Species	Cod 147; Had 52; Her 24	Cod 1,218; Had 497; Her 104	Cap 510; P Cod 148; Cod 95	Cap 442; P Cod 348; Cod 236	Cap 479; Cod 222; Had 167	Cap 1,140; Cod 510; Had 236	Cap 773; Cod 657; Had 148
Main Countries	USSR 172; UK 90	Nor 1,040; USSR 745; UK 107	Nor 485; USSR 194; UK 114	Nor 541; USSR 496; UK 64	Nor 555; USSR 470; UK 53	USSR 1,247; Nor 664; UK 87	Nor 832; USSR 754; UK 96

Svalbard Waters (ICES sub-area IIb)

	1953	1962	1971	1972	1973	1974	1975
TOTAL		70	248	262	100	359	598
Main Species	Cod	Cod 59	Cap 218; Cod 149	Cod 72; Red 64; Gre 60	Cap 39; Cod 19; Red 18	Cap 196; Cod 89; Red 29	Had 244; Cap 244; Red 42
Main Countries	UK; W Ger	USSR 55; UK 9	USSR 149; Nor 73; Pol 26	USSR 152; Nor 85; UK 9	Nor 71; USSR 15; UK 9	Nor 207; USSR 121; W Ger 16	USSR 255; Nor 230; W Ger 45

Norwegian Sea (ICES sub-area IIa)

	1953	1962	1971	1972	1973	1974	1975
TOTAL		777	939	1,679	1,790	600	654
Main Species	Her; Cod; Pol	Her 364; Cod 209; Pol 100	Cap 506; Cod 199; Pol 109	Cap 878; Cod 424; Pol 180	Cod 1,108; Pol 382; Had 173	Cod 233; Pol 184; Cap 48	Cod 243; Pol 226; Red 131
Main Countries	Nor; W Ger; UK	Nor 681; USSR 44; UK 35	Nor 644; Ice 212; UK 44	Nor 1,470; UK 75; W Ger 49	Nor 1,698; W Ger 33; E Ger 29	Nor 504; E Ger 39; W Ger 27	Nor 643; W Ger 59; E Ger 49

Iceland (ICES sub-area Va)

	1953	1962	1971	1972	1973	1974	1975
TOTAL		965	1,365	1,018	987	1,137	1,141
Main Species	Cod; Red; Her	Her 515; Cod 157; Had 95	Cod 651; Cap 386; Pol 120	Cod 453; Cap 183; Pol 134	Cap 399; Cod 277; Pol 108	Cap 442; Cod 380; Pol 111	Cap 462; Cod 375; Pol 98
Main Countries	Ice; UK; W Ger	Ice 435; UK 242; Nor 217	Ice 818; UK 204; W Ger 158	Ice 627; UK 210; W Ger 125	Ice 675; UK 185; W Ger 94	Ice 858; UK 155; W Ger 92	Ice 888; UK 142; W Ger 68

East Greenland (ICES sub-area XIV)

	1953	1962	1971	1972	1973	1974	1975
TOTAL	Not reported separately		47	68	49	33	53
Main Species		Red; Cod	Cod 27; Red 17	Cod 32; Red 20	Gre 27; Cod 13	Gre 13; Red 12	Red 25; Gre 20; Cod 6
Main Countries		W Ger	W Ger 41; Pol 9; Ice 7	W Ger 44; Pol 9; Ice 7	W Ger 29; E Ger 9; Ice 9	E Ger 14; Ice 10; W Ger 4	E Ger 23; USSR 11; W Ger 7

Table: Northwest Atlantic fisheries catches by ICNAF sub-area (thousands of tons). The six data columns are unlabelled in this excerpt and are numbered (1)–(6); the first column gives amounts "Not reported separately".

Baffin Island (ICNAF sub-area o)

	Not reported separately	(1)	(2)	(3)	(4)	(5)	(6)
TOTAL		16	3	4	2	2	
Main Species			Rng 2, Gre 1				
Main Countries			Rng, Gre, USSR	Gre	USSR, Den		

West of SW Greenland (ICNAF sub-area 1)

	Not reported separately	(1)	(2)	(3)	(4)	(5)	(6)
TOTAL		530	150	139	105	111	142
Main Species	Cod 223, Red 205	Cod 451, Shr 60	Cod 121, Shr 7	Cod 111, Shr 10, Gre 4	Cod 63, Gre 13, Shr 7	Cod 48, Shr 13, Gre 18	Cod 48, Shr 38, Gre 23
Main Countries	Por 55, Den 53, UK 35	W Ger 192, Den 93, Por 92	Den 55, Nor 43, W Ger 23	Den 53, Nor 33, Spa 20	Den 48, USSR 19, Por 10	Den 60, USSR 18, W Ger 10	Den 60, USSR 37, W Ger 16

Labrador coast (ICNAF sub-area 2)

	Not reported separately	(1)	(2)	(3)	(4)	(5)	(6)
TOTAL		266	246	220	159	255	286
Main Species	Cod 109	Cod 255, Rng	Cod 163, Cap 57	Cod 163, Cap 18, Gre 12	Cod 60, Cap 58, Gre 14	Cap 125, Cod 85, Red 16	Cap 145, Cod 89, Red 15
Main Countries	Por 59, Fra 27, Spa 15	USSR 68, Por 63, Pol 57	USSR 165, Pol 34, Por 21	USSR 133, Pol 24, Por 20	USSR 106, Pol 15, W Ger 14	USSR 126, Pol 37, W Ger 34	USSR 188, Pol 34, W Ger 25

	Not reported separately	(1)	(2)	(3)	(4)	(5)	(6)
TOTAL, ALL AREAS LISTED ABOVE	2,418[a]	4,211	4,578	4,411	4,451	4,791	4,845
TOTAL, ALL WORLD SEA AREAS	24,100	40,050	59,911	55,132	55,282	58,898	57,989

[a] Excludes the Soviet catch, so this total should be higher. The USSR probably fished only in the Barents and Norwegian Seas, but its catch there in 1955 was nearly 900,000 tons. The omission of a figure for East Greenland is not significant.

Abbreviations

Cap	Capelin, *Mallotus villosus*	Den	Denmark (includes Faeroes and Greenland)
Cod	*Gadus morhua*	E. Ger	German Democratic Republic
Gre	Greenland halibut, *Reinhardtius hippoglossoides*	Fra	France
Had	Haddock, *Melanogrammus aeglefinus*	Ice	Iceland
Her	Herring, *Clupea harengus harengus*	Nor	Norway
P Cod	Polar cod, *Boreogadus saida*	Pol	Poland
Pol	Pollock or saithe, *Pollachius virens*	Por	Portugal
Red	Redfish, *Sebastes marinus*	Spa	Spain
Rng	Roundnose grenadier, *Macrourus rupestris*	USSR	USSR
Shr	Shrimp, *Pandalus* sp.	UK	UK
		W Ger	Federal Republic of Germany

(Sources: *International Commission for the Northwest Atlantic Fisheries; Statistical Bulletin, Dartmouth, Nova Scotia; Bulletin Statistique des Pêches Maritimes, Charlottenlund, Conseil International pour l'Exploration de la Mer; Yearbook of Fishery Statistics, Rome, Food and Agriculture Organization of the United Nations.*)

Table 34. Northern North Atlantic. Whales and harp seals caught by species and grounds, 1955–75 (number of animals)

Grounds and species	1955	1960	1965	1970	1971	1972	1973	1974	1975
				Whales					
Norway									
Fin	210	128	106	44	37	—	—	—	
Sperm	44	84	27	51	62	—	—	—	
TOTAL[a]	266	214	133	95	99	—	—	—	
Small whales[b]	4,588	4,035	3,294	3,143	2,645	2,695	2,059	1,827	
Iceland									
Fin	236	160	288	272	208	238	267	285	
Sei	134	42	74	44	240	132	139	9	
Sperm	20	177	70	61	106	76	47	71	
TOTAL[a]	400	379	432	377	554	446	453	365	
Small whales[b]	—	—	—	—	—	35	122	86	
Greenland									
TOTAL	35	1	—	1	5	5	12	13	
Small whales[b]	—	591	2,041	3,218	2,478	2,154	2,320	1,891	
Faeroe Islands									
TOTAL	161	—	16	—	—	—	—	—	
Small whales[b]	n.a.	1,702	1,776	448	1,116	511	1,051	677	
Canada (Newfoundland, Labrador, Baffin Bay)									
Fin	2	1	6	460	301	265	—	—	
TOTAL[a]	2	2	9	463	317	267			
Small whales[b]	6,626[c]	1,583	1,549	331	285	155	393	427	
Total Northern North Atlantic[d]									
Fin	550	289	410	776	546	503	267	285	
Sei	158	44	74	45	241	132	139	9	
Sperm	142	261	103	112	168	77	47	71	
TOTAL[a]	864	596	590	936	995	718	465	378	
Small whales[b]	11,214	7,320	8,660	7,140	6,524	5,553	5,945	4,908	
				Harp Seals					
Canada									
Maritime	n.a.	n.a.	79,936	43,139	28,561 ⎫				
Quebec	n.a.	n.a.	10,921	10,475	32,086 ⎬ 76,583	65,542	92,050	114,202	
Newfoundland	n.a.	n.a.	76,239	88,681	72,013 ⎭				
Northwest Atlantic									
Greenland[e]	n.a.	n.a.	9,251	36,000	33,000	21,488	54,255	40,866	n.a.
Norway	n.a.	n.a.	67,157	115,200	98,639	53,300	58,290	55,585	60,161
Northeast Atlantic									
Greenland[e]	n.a.	n.a.	1,882	16,000	16,000	19,271		14,057	n.a.
Norway	n.a.	n.a.	30,608	31,060	19,781	27,620	24,179	22,079	15,767

n.a. = not available
[a] Totals include other species not listed.
[b] Includes minke, bottlenose, pilot, white (beluga), killer, Baird's beaked, narwhal, and common porpoises.
[c] 6,612 pilot whales taken in 1955.
[d] Entire take by shore-based catcher boats. Canadian take of 'small whales' includes Eskimo subsistence harvest in Baffin Bay area.
[e] Greenland data include some hooded seals.
[f] Included with West Greenland total.
(*Sources: International Whaling Statistics*, Oslo, Committee for Whaling Statistics; *Yearbook of Fisheries Statistics*, Rome, Food and Agriculture Organization of the United Nations.)

International agreements

Leaving aside broad multi-lateral agreements which pertain to all marine areas, such as the International Convention for the Safety of Life at Sea, there are two international agreements which refer specifically to this area. Both relate to fisheries. In 1946 a 'Convention for the regulation of the meshes of fishing nets and the size limits of fish', covering the northeast Atlantic, was signed by most of the countries fishing in the area, and adhered to later by the USSR and West Germany. This convention was broadened in scope to become the North East Atlantic Fisheries Convention (NEAFC), which came into force in 1963 and is administered by a permanent commission of the same name. The scientific work on which these conventions rest was, and is, performed by the Conseil Permanent International pour l'Exploration de la Mer, which was set up in Copenhagen as early as 1902, but is a scientific body and has never had any sort of executive role.

Parallel to NEAFC is ICNAF, the International Convention (and Commission) for the Northwest Atlantic Fisheries, which came into force in 1950. There is a substantial overlap in membership, but Canada and the USA belong only to ICNAF. These two on-going commissions now exercise considerable influence in the management of the fishery. Although their recommendations are not binding on governments, there is clearly strong moral pressure on governments to accept internationally agreed restrictions. Since 1973 most member countries have agreed to allow inspection of their catches by protection officers of other member countries – a considerable step forward.

These are the two major instances of international agreements covering the northern North Atlantic. Whaling in the area is subject to the provisions of the International Convention for the Regulation of Whaling signed in 1946. There are also certain bilateral agreements, such as that between the USSR and Norway concerning exploitation of the harp seal populations off Jan Mayen and in the White Sea (the more important Labrador population is superintended by ICNAF). Another bilateral agreement is that between the UK and Iceland, reached in 1973, on fishing rights in Iceland waters. Unhappily this agreement did not prevent the 'cod wars' of the mid-1970s. This underlines the desirability of wider accord among fishing nations as to what can and cannot be done. There is bound to be more and more pressure on the resources of the sea, and it is in the interests of all to agree on efficient and effective management practices.

Seabed exploitation

Some of the problems posed by international regulation of fisheries arise equally in connection with exploitation of the seabed. There is as yet, however, no such development in this area. But offshore oil exploration is now being undertaken in three general locations: off the Labrador coast, off the West Greenland coast, and in Svalbard waters. If oil is found there, as seems likely, drilling platforms and other structures will be built. The technical problems will be considerable, since floating ice will be present at least part of the year, and in the Labrador and West Greenland regions some of the ice will be in the form of icebergs, which are capable of doing much more damage than sea ice. But the drive to find new sources of energy will no doubt be strong enough to overcome these difficulties.

The North Pacific

The portion of the North Pacific covered in this chapter is the area north of a line extending from Dixon Entrance off the North American coast (approximately latitude 54° 30' N) westward to a point offshore from the eastern end of the Aleutian Island chain at the intersection of 170° W with 50° N, continuing westward along 50° N (see Map 21). This includes all of the Bering Sea, the Sea of Okhotsk, the North Pacific off the Aleutian Islands, and the Gulf of Alaska (International North Pacific Fisheries Commission statistical areas Shumagin, Chirikof, Kodiak, Yakutat, and Southeastern). The Chukchi Sea, north of the Bering Sea, is included with the Arctic Ocean in this book.

The area covered by this region is about 6 million sq km. The continental shelf is narrow off southeastern Alaska (approximately 40–50 km out to the 200 m contour), widens up to 100 km off Kodiak Island and narrows again westward off Unimak Pass. It underlies about half of the Bering Sea and most of the Sea of Okhotsk. The total continental shelf area in this region is about 2 million sq km. The Gulf of Alaska areas have mountainous coasts with numerous islands and long inlets which contrast sharply with the gentle sloping lands, open shores, relatively shallow waters, and general Arctic conditions of the Bering Sea and the Sea of Okhotsk.

The current systems are shown on Map 21. Those circulating in the Gulf of Alaska and up into Bering Strait are chiefly warmer, and those coming down the Siberian shore and round the deep basin of the southwest Bering Sea are colder. The cyclonic flow in the Sea of Okhotsk is warm.

A band of high phytoplankton production extends from the southeastern Alaska region along the coast of the Gulf of Alaska and the Aleutian Islands. The open oceanic part of the North Pacific is much more productive than the tropical and sub-tropical areas to the south. Extremely

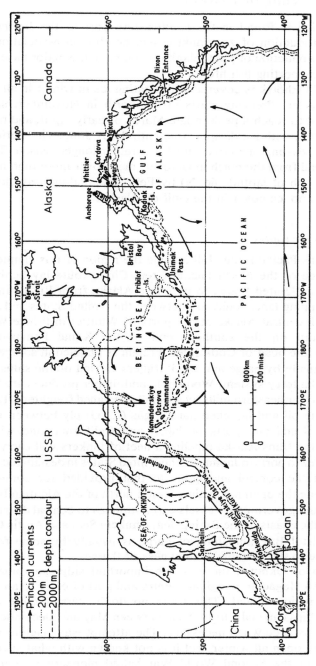

Map 21. North Pacific fishing grounds

high production is found in the southeastern part of the Bering Sea. Secondary production of zooplankton and benthos in association with the primary production of phytoplankton comprise the major base of the food chains supporting fish life.

The Sea of Okhotsk is covered with floating ice in winter in all but its southeast quarter. Freezing starts in the north in November, spreads southeastward to reach a peak in late March or early April, and retreats to the northwest with ice disappearing entirely by June. The Bering Sea similarly has seasonal ice cover but for a rather longer season. The ice starts to spread from the north in October, may cover more than half the sea (i.e. to south of latitude 60° N) by February, and entirely clears by July. There is no ice cover in the Gulf of Alaska.

Shipping and air routes

The main trans-Pacific routes fall outside the region of this chapter, but the importance of the North Pacific Great Circle route lay behind the decision by the United States to purchase Alaska from Russia in 1867. The main shipping lanes today are between the United States and Japan, entering south-central Alaska at the ports of Whittier in Prince William Sound, Seward on the southern end of the Kenai Peninsula and Anchorage at the head of Cook Inlet. Freight is handled in southeastern Alaska principally by barge and truck trailers carried on the ships of the State of Alaska ferry system. Wood pulp and timber products are hauled to Japan from southeastern mills by specially designed freighters. Tankers carrying crude oil and liquefied natural gas (LNG) ply between marine terminals at Kenai and ports in Japan and California, and with full operation of the trans-Alaska pipeline a fleet of tankers will carry crude oil from Valdez to ports in Japan and the United States in designated tanker lanes with traffic control analogous to that provided for commercial aircraft. During the open navigation season north of the Aleutian Islands, commercial and government vessels carry supplies to central distribution points along the coasts of the Bering Sea, Chukchi Sea, and Arctic Ocean. North of Bristol Bay unloading of cargo is usually by some form of lighterage due to the extreme shallowness of the sea.

In the Sea of Okhotsk there is one important shipping route from Vladivostok to Magadan, a regional centre and port of entry to an important mining region inland. This route no doubt carries quite a high priority and ships can make the voyage between May and December, with icebreaker assistance when necessary. The attempt was made in 1975 to keep the route open all winter, but it is not known with what success.

The onset of the Second World War halted pioneering attempts to establish intercontinental commercial air service between North America

Table 35. Northern North Pacific. Whales caught and fur seals harvested, by species and grounds, 1955–75 (number of animals)

Grounds and species	1955	1960	1965	1970	1971	1972	1973	1974	1975
Whales									
Japan: factory-ship fleets (North Pacific, excluding Japan grounds and Bering Sea)									
Fin	1,360	1,393	1,406	518	542	426	256	216	
Sei	21	203	1,398	3,235	2,431	2,041	1,710	1,195	
Sperm	1,084	1,800	2,460	2,700	1,803	1,567	1,803	1,803	
TOTAL[a]	2,652	3,466	5,353	6,453	4,874	4,039	3,771	3,731	
USSR: factory-ship fleets (primarily Kamchatka grounds)									
Fin	79	128	1,492	412	190	250	161	173	
Sei	28	59	695	782	296	71	103	42	
Sperm	996	2,228	8,196	8,585	5,525	1,736	4,329	3,963	
TOTAL[a]	1,115	2,472	10,698	9,845	6,649	2,128	5,249	4,832	
USSR: shore-based (Kamchatka and Kuril Islands)									
Fin	219	265	—	—	—	—	—	—	
Sei	128	140	—	—	—	—	—	—	
Sperm	1,494	1,487	—	—	—	—	—	—	
TOTAL[a]	1,901	1,909	—	146	150	182	178	—	
Small whales	44	97	—	—	—	—	—	—	
Total northern North Pacific[b]									
Fin	1,658	1,786	2,898	930	732	676	417	389	
Sei	177	402	2,093	4,017	2,727	2,112	1,813	1,237	
Sperm	3,574	5,515	10,656	11,285	7,328	3,303	6,132	5,766	
TOTAL[a]	5,668	7,847	16,051	16,444	11,673	6,349	9,198	8,563	
Small whales	44	97	—	—	—	—	—	—	
North Pacific Fur Seal									
USSR: Commander Islands (Komandorskiye Ostrova)			20,019	16,306	15,372	12,690	8,582	4,227	4,200
USA: Pribilof Islands	65,453	40,616	51,000	42,179	31,824	37,221	28,582	33,027	28,849

[a] Total includes species not itemized.
[b] Does not include unreported subsistence harvest by St Lawrence Island Eskimo. Amount is minor. Most of Alaska's aboriginal harvest is in Arctic (see Table 37).
(*Sources*: *International Whaling Statistics*, Oslo, Committee for Whaling Statistics; *Yearbook of Fisheries Statistics*, Rome, Food and Agriculture Organization of the United Nations.)

and the Far East via a route which required fuel stops in northern British Columbia, southeastern Alaska, Yukon Territory, Fairbanks, Nome, the Aleutian Islands, the Kuril Islands, and Japan. Technological advances made possible revision to a route with stops in southeast Alaska, Fairbanks, and Tokyo in 1957, and further advances in aircraft range permitted

Table 36. Commercial fisheries nominal catch (live weight equivalent of fish landed) in northern North Pacific, by fishing area, main species, and main countries, 1955–75 (thousand tonnes)

	1955	1965	1970	1971	1972	1973	1974	1975
I *Gulf of Alaska*[a]								
Japan								
Rockfish (primarily ocean perch)	—	43	45	49	53	54	41	34
Pollock	—	3	9	9	14	7	30	10
Sablefish	—	2	24	25	36	27	24	18
Flounder	—	5	4	2	8	19	7	2
Other ground fish	—	—	3	3	2	7	10	9
TOTAL	—	53	85	88	113	114	112	73
USSR								
Rockfish	n.a.	n.a.	*	30	24	4	17	10
Pollock	n.a.	n.a.	*	tr.	20	30	31	38
Atka mackerel	n.a.	n.a.	*	*	*	9	18	20
Other ground fish	n.a.	n.a.	9	1	25	10	12	11
Shrimp	—	7	4	5	2	2	n.a.	n.a.
Other fish	n.a.	53	85	88	113	114	112	73
TOTAL	n.a.	400	98	124	184	169	190	152
USA								
Halibut	12	18	15	9	14	8	7	9
Other ground fish	2	tr.	tr.	1	1	1	1	1
Salmon	76	66	98	80	72	51	45	41
Herring	29	12	7	5	8	10	18	17
Shrimp	1	6	34	43	38	49	53	45
Scallops	—	—	1	1	1	1	tr.	tr.
King crab	3	50	14	14	16	17	17	16
Tanner (snow) crab	tr.	tr.	6	6	14	28	27	18
Dungeness crab	2	4	4	2	3	2	2	1
TOTAL	125	156	179	161	167	166	170	148
Canada								
Halibut	10	20	11	9	8	6	3	2
Other ground fish	—	—	tr.	tr.	tr.	tr.	tr.	tr.
Republic of Korea								
Sablefish	—	—	—	—	—	1	3	2
Pollock	—	—	tr.	tr.	1	1	*	*
Other ground fish	—	—	—	—	3	2	3	2
Poland								
Ground fish	—	—	—	—	—	tr.	tr.	4
II *Bering Sea – Western Aleutians*[b]								
Japan								
Sablefish	—	9	11	16	16	7	7	6
Flatfish (other than halibut)	n.a.	39	124	184	177	180	226	138
Pacific cod	n.a.	20	75	46	40	41	48	35
Ocean perch	n.a.	17	23	25	14	13	30	12
Pollock	n.a.	231	1,232	1,515	1,652	1,472	1,253	1,053
Other grouNd fish	15	37	9	13	13	10	10	10
Salmon (mothership fishing area)[c]	64	45	36	37	35	34	33	40
Herring	—	36	28	23	6	2	6	2

Table 36. (cont.)

	1955	1965	1970	1971	1972	1973	1974	1975
Shrimp	—	9	—	—	—	—	—	—
Octopus squid	—	—	—	—	1	—	—	4
King crab	3	12	5	9	8	6	5	2
Tanner (snow) crab	n.a.	n.a.	n.a.	28	28	21	25	16
TOTAL	82	455	1,549	1,903	1,995	1,820	1,643	1,318
USSR								
Rockfish	—	n.a.	53	7	25	4	33	39
Pollock[d]	—	—	45	1,039	853	1,565	1,727	1,920
Flounder	—	n.a.	115	143	61	21	39	50
Other ground fish	—	n.a.	19	7	104	20	16	25
Salmon (USSR Bering Sea – Kamchatka areas)	119	30	18	24	8	17	18	31
Herring	—	10	117	23	54	34	20	19
King crab	—	n.a.	n.a.	20	15	19	18	16
TOTAL	119	210	367	1,263	1,120	1,680	1,866	2,100
USA								
Salmon	17	59	59	34	14	11	15	21
Halibut	tr.	tr.	tr.	tr.	tr.	tr.	1	tr.
King crab	1	10	10	18	17	18	23	24
Tanner crab	—	—	1	tr.	tr.	tr.	2	3
TOTAL	18	69	70	52	31	29	41	48
Canada								
Halibut	tr.	tr.	1	1	tr.	tr.	1	tr.
Republic of Korea								
Pollock	—	—	5	10	9	3	3	3
III *Sea of Okhotsk and adjacent areas*[e]								
Japan								
King crab	—	—	—	9	8	6	5	2
Tanner crab	—	—	—	25	28	21	25	16
USSR								
Salmon	118	64	21	54	23	61	31	72
Herring[f]	106	250	280	282	270	323	285	314
King crab	—	—	—	20	18	19	18	16

n.a. = not available.

* Included with 'other' if any.

tr. = trace (less than 500 tonnes).

[a] *INPFC areas*: Shumagin, Chirikof, Kodiak, Yakutat, and Southeastern.

[b] *INPFC Bering Sea region*: all Bering Sea waters south of Chukchi Sea and north and south sides of Aleutian Islands between longitudes 170° W and 170° E.

[c] *Japan mothership area*: Bering Sea and northern North Pacific between latitudes 52° N and 46° N and west of longitude 175° W.

[d] Total of pollock catch in eastern Bering Sea reported in North Pacific Fisheries Management Council sources and Alaska pollock catch reported for FAO Northwest Pacific region.

[e] USSR salmon fisheries includes areas off Sakhalin and Kuril Islands. Portion of USSR, Japanese, and Korean ground fish fisheries within this sub-region not reported.

[f] Estimate based on data in Gulland and FAO *Yearbooks*.

(*Sources*: *Statistical Yearbook*, Vancouver, International North Pacific Fisheries Commission; *Yearbook of Fisheries Statistics*, Rome, Food and Agriculture Organization of the United Nations; North Pacific Fisheries Management Council (draft management plans), Anchorage, Alaska; *Alaska Catch and Production, Commercial Fisheries Statistics* (annual), Juneau, Alaska, Department of Fish and Game: J. A. Gulland, ed. *The fish resources of the oceans*, London, 1971.)

one-stop flights (at Fairbanks) between Tokyo and Seattle or New York. Anchorage, however, has become the mid-point on nine international and domestic airline flights connecting Tokyo and the Far East with North American and European centres.

Fisheries

The fish and aquatic mammals of the region have long provided the main subsistence base for the indigenous people and they continue to make important contributions to their subsistence today. The first commercial exploitations of these resources were of sea otter and fur seal by the Russians commencing in the eighteenth century, and of whales and walrus by the USA in the nineteenth. The sea otter was thought to be extinct by 1867 when Alaska was sold to the United States by Russia, but survivors were discovered in this century. Although their numbers have increased, they continue to be a protected species. Heavy pelagic sealing almost exterminated the fur seal by the end of the century, but the resource made a remarkable recovery under the protection and management provided by the 1911 treaty between the United States, Canada, Russia, and Japan. The Pribilof herds, for example, increased from fewer than 150,000 animals at the treaty's inception to about 1,750,000 currently. Harvests are made only of surplus males on the rookeries of the Commander Islands (Komandorskiye Ostrova) and Pribilof Islands, the average annual number of skins taken for the five-year period 1971–5 being 9,014 by the USSR and 31,881 by the USA.

Whales were severely depleted and the major commercial operations terminated early in the present century, and the last Alaskan commercial land-based stations were abandoned by the 1920s. Japan resumed whaling in the western Bering Sea in the 1950s, the number of whales taken by motherships increasing from 2,652 in 1955 to 7,548 in 1968 and falling to 3,731 by 1974. The USSR entered two new factory ships into its North Pacific whale fishery in 1963, and thereby substantially increased the catch. The number of whales killed in 1959 was 1,881 and rose to 10,059 by 1963 and to 12,386 by 1966. For the period 1963–70 the annual Soviet harvest averaged 10,860, but had fallen to 4,832 by 1974 (Table 35).

Of the commercial fisheries off Alaska, the oldest appears to be a long-line cod fishery by US schooners starting in 1864. The halibut fishery started in the Gulf of Alaska in 1890 by Canadian and US fishermen and following serious depletion of the resource, the harvest both in the gulf and south-eastern Bering Sea has been strictly controlled by the International Pacific Halibut Commission. Japan entered a trawl fishery in the Bering Sea in the 1930s for flatfish and pollock. This fishery was re-established in 1954 and Soviet bottomfish and herring winter fisheries appeared in 1958 and were extended to the Gulf of Alaska in 1962. By the mid-1960s more than 70 per cent of the Soviet landings from its North Pacific factory-ship operations came from waters off Alaska. Before these dates, the bottom-fish catch was almost entirely by Canadian and United States fishermen seeking halibut, cod, and sablefish. Since then the foreign fisheries have included fleets from the Republic of Korea and Poland in addition to Japan and USSR. These foreign trawl fisheries take Pacific halibut, sablefish, six species of flounder, Pacific ocean perch, Alaska pollock, Atka and jack mackerel, Pacific hake, and several varieties of rockfish. They also take herring with gillnets in the southeastern Bering Sea (Table 36).

The Soviet fisheries in the Sea of Okhotsk and the western Bering Sea include the commercial take of a similar range of species. Before 1958 these fisheries had been confined to coastal areas, mainly for herring and salmon. Since the mid-1950s the effort has been greatly expanded offshore with investment in fleets of modern fishing, processing, and support vessels, and reorganization of shore-based processing centres to meet this transformation. Japanese trawlers from the Pacific side of Hokkaido also fish in the Sea of Okhotsk and the western Bering Sea for a wide variety of bottomfish.

Traditionally the United States has fished king crab (genus *Paralithodes*) in the Gulf of Alaska and for short periods in the eastern Bering Sea, while Japan and the USSR have fished in the Okhotsk Sea and eastern Bering Sea. Evidence of declines in king-crab stocks in all areas have turned the crabbing effort towards snow or Tanner crab (genus *Chionecetes*). The Dungeness crab (*Cancer magister*) also occurs in the region and is har-vested by US fishermen in the Gulf of Alaska and southeastern Alaska. The prawn and shrimp resources of the region have been harvested principally in the Bering Sea by the Japanese fleet and in the Gulf of Alaska by Soviet and US fishermen. Due to overfishing, Japanese catches dropped from 21,000 tonnes in 1964 to 9,000 tonnes in 1965 and continued to fall until the fishery was discontinued after 1968. A significant commercial fishery for weathervane scallop (*Patinopecten caurinus*) was developed in 1968 in the Gulf of Alaska by US fishermen. Since the 1964 earthquake, which altered many Alaska beaches, the harvest of clams in the Gulf of Alaska has been insignificant and development elsewhere has been hindered by fear of paralytic shellfish poisoning.

The five species of Pacific salmon (genus *Oncorhynchus*) have long been the most important commercial fisheries in the North Pacific and Bering Sea. Salmon originate in the spawning areas drained by streams and rivers along the North American, Siberian, and northern Japanese coasts, migrate to sea where they spend varying periods of time growing to maturity, then return to their parent nursery areas to spawn and die. The United States and Canadian commercial salmon fisheries grew rapidly during the 1880s, reached peaks in the 1930s and have since declined drastically. The salmon resources continue to provide important subsistence harvests and the basis of a growing sports fishery in southcentral and southeast Alaska. The Soviet salmon fisheries appear to have been insignificant until the 1920s, after which they followed a pattern similar to the North American with peak production in the 1940s. A wide variety of gear has been used in these fisheries (traps, gillnets, seines, and lines) and the USSR, Canada, and the United States have limited their fisheries to the interception of returning mature salmon within their respective territorial waters. The Japanese experience has been otherwise.

In addition to the harvest of salmon off their own northern islands in which they had been engaged since the sixteenth century, the Japanese established shore-based fisheries in north Sakhalin, the north Kurils, the Sea of Okhotsk, and the Kamchatka and Siberian coasts of the Bering Sea under provisions of the Japan–Russia Fishery Convention of 1907. In the 1930s a mothership gillnet fishery was operated outside the territorial seas off Kamchatka. Between 1910 and 1944 Japanese fishermen caught more salmon than the fishermen of the other nations in the region. This lead was lost during the Second World War, but within a decade of its end Japan had re-established its ranking as the leading salmon fishing nation, primarily by extensive fishing on the high seas.

The several Japanese high-seas salmon fisheries have been the focus of much of the controversy over fishing in the North Pacific and the target of international attempts to manage the total resource. In 1952 Japan launched a mothership gillnet fishery and a land-based driftnet fishery, and between 1952 and 1971 operated a land-based longline, high-seas fishery in the North Pacific. Between 1955 and 1958 it also operated a mothership salmon fishery in the Okhotsk Sea. Only the areas fished by the North Pacific mothership fleet since 1959 fall within the boundaries of the region of this chapter and, although North American and Siberian salmon are frequently taken in substantial quantities by the other high-seas fisheries, discussion will be limited to this single fishery. As prescribed by provisions of the 1952 convention with the United States and Canada and the 1957 treaty with the USSR, the fishery operates within an area of 2.5 million sq km outside Siberian territorial waters between longitudes 160° E and 175° W and between approximately latitudes 46° N and 61° N. The size of the fleet stabilized between 1972 and 1977 at 10 motherships ranging from 7,700 to 14,000 gross registered tons accompanied by 332 catcher-boats of 75–100 tons.

The United States bases its fight to outlaw or limit more stringently the Japanese high-seas fisheries on several points derived from intensive tagging programs carried out between 1956 and 1975, reports of observers on Japanese ships, and theories on salmon migration and behaviour on the high seas. The first is that taking salmon while the various races are intermingled cannot be managed scientifically and attempts of the other nations to do so within their territorial waters are thwarted. The twenty-year tagging studies, for example, indicate that 23 per cent of the total western Alaska sockeye salmon catch was by the Japanese mothership fishery and for other species this take has fluctuated in an unpredictable manner from year to year, in several years exceeding half of the total catch. The second major point is that the fishery is wasteful on several counts. Because the high-sea catch is of immature salmon, there is considerable loss of body weight as compared with a catch of the same fish at maturity (e.g. the studies indicate that the tonnage of sockeye taken pelagically could have been increased 1.76 times if taken in western Alaska and that chinooks could have increased more than threefold in weight), and loss of immature salmon from drop-outs of dead fish and escape of injured fish is about twice the loss for mature salmon. In addition, observers on Japanese motherships and research vessels estimate that 11,800 Dall porpoise and 235,000 sea birds are killed annually owing to use of high-seas gillnets.

International agreements

The first major international agreement in the region was the treaty of 1911 between the United States, Canada, Japan, and Russia for the conservation and management of the North Pacific fur seal resource. The pelagic taking of fur seals was outlawed and the management and harvest of the herds on the Commander Islands (Komandorskiye Ostrova) in Russia and the Pribilof Islands in Alaska became the responsibility of Russia and the United States, respectively, with the net proceeds of the sale of skins allocated among the parties to the treaty. Following suspension during the Second World War, the arrangement was reinstated by the Interim Convention for the Protection of North Pacific Fur Seal, February 1957. The most recent agreement was negotiated in 1969 to expire in 1987.

The International Convention for the High Seas Fisheries of the North Pacific Ocean, signed by Canada, United States, and Japan in May 1952, established provisional lines at longitudes 175° W in the Bering Sea and 175° 20′ W in the North Pacific east of which Japan agreed to abstain from fishing salmon. Canada agreed to abstain from fishing salmon east of the provisional line in the Bering Sea. Formal extension of the abstention provisions beyond the initial five years has never been made, as agreement could not be reached, but Japan is still obligated to observe the abstention lines. In December 1972 the United States and the Republic of Korea entered into a similar agreement with Korea abstaining from fishing salmon east of the 175° W line. Under the 1957 Fisheries Treaty for the northwest Pacific Ocean between Japan and the USSR, annual catch quotas are set for Japanese salmon fisheries on the high seas west of 175° W. Since 1958 Japan has been excluded from the Okhotsk Sea fisheries, the mothership fishing area has been moved farther off the east coast of Kamchatka, and minimum distances between gillnet sets have been increased.

Halibut fisheries have been under joint Canadian–United States management since 1924. Currently, they are subject to provisions of the Convention between the United States of America and Canada for the Preservation of the Halibut Fishery of the Northern Pacific Ocean and Bering Sea (1937 as amended in 1953) and the conventions cited above. The Convention waters are divided into areas within which fishing seasons are established, limits placed on size of fish and quantity of catch, gear regulated, and nursery grounds designated.

In 1964 the United States ratified the 1958 Continental Shelf Convention and bilateral agreements were concluded with Japan and the USSR applying quotas and minimum size limits to king crab catches, and in 1969 to Tanner (snow) crab. Modifications every two years progressively lowered catch quotas and prohibited use of tangle nets and retention of females and undersized crab.

Following United States establishment in 1966 of a 9-mile continuous fishery zone adjacent to the 3-mile territorial sea, agreements were negotiated by the United States with Japan, the USSR, and Canada between 1967 and 1970 providing for abstentions and/or regulated participation in specific fisheries. In 1973 the United States agreed with the USSR, Japan, and the Republic of Korea on procedures to reduce conflict over fixed and mobile gear and to resolve claims when conflicts occur.

Whaling is subject to the provisions of the International Convention for the Regulation of Whaling of 1946. Effective as from 1 March 1977, the United States also amended its Marine Mammal Protection Act of 1972 to extend its protection of marine mammals, including whales, to include all waters in the 200-mile fisheries zone established in the Fishery Conservation and Management Act of 1976, to be discussed below. Because of the continued subsistence whale fishery of the Eskimo people along the Arctic Ocean and Bering Sea and their use of other sea mammals, special provisions are made for the pursuit of their traditional hunting activities.

The Fishery Conservation and Management Act of 1976 was enacted upon findings by the United States Congress that 'activities of massive foreign fishing fleets' threatened survival of fish off the coast of the United States and that 'international fishery agreements have not been effective in preventing or terminating the overfishing of these valuable resources.' The United States asserts the right to exercise exclusive management authority over all fisheries within a fishery conservation zone measured 200 nautical miles from the baseline from which the territorial sea is measured, all anadromous species spawned in waters of the United States throughout their migration range (except when in territorial waters or fishery conservation zones of other nations), and all continental shelf resources of the United States beyond the conservation zone. Foreign fishing is allowed from that portion of the optimum yield of each fishery which will not be harvested by vessels of the United States. The fisheries of the Arctic Ocean, the Bering Sea, and the Pacific Ocean seaward of Alaska are under the management of the North Pacific Fishery Management Council at Anchorage.

Total allowable catches and allocations between United States and foreign fishing fleets were made for all principal species except salmon for the 1977 season. The North Pacific fisheries treaty of 1952 remains in effect until one year from the day on which one of the three contracting nations gives notice of intention of termination of the Convention, and this was not done. Consequently, Japan continued to fish for salmon on the high seas west of 175° W pending the outcome of talks between the United States and Japan. Although the 1911 fur-seal treaty presents the model for a solution of the pelagic taking of salmon, the nations involved have made no apparent moves in that direction.

Seabed exploitation

For the near future, seabed resource exploitation appears promising only in offshore oil and gas developments. Outer continental shelf oil and gas leases were let by the United States government in the eastern Gulf of Alaska near Yakutat in 1976 and others are scheduled in the eastern Gulf near Cordova (1979), lower Cook Inset (1977 and 1980), western Gulf off Kodiak Island (1977 and 1980), Aleutian Island shelf (1980), St George Basin in the southeastern Bering Sea (1980), Bristol Bay (1980), and Norton Basin in the northeastern Bering Sea (1979). Exploratory activities were launched in the northern Gulf and St George Basin in 1976 and shore facilities were placed at Yakutat and Seward.

No international problems appear to be directly posed by these offshore oil and gas activities, but the potential petroleum provinces do occur in rich marine environments and could create serious conflicts with existing domestic and foreign fisheries. The technical engineering problems are enormous. In the Gulf of Alaska the platforms, underwater storage, and other structures must be able to withstand fierce storms and icing conditions and the hazard of frequent strong earthquake shock, and in the Bering Sea must cope with the problems of sea ice.

The Arctic Ocean

The Arctic Ocean is the body of water about the North Pole, enclosed by the northern coasts of America, Asia, and Greenland. It will be taken here to include the Beaufort, Chukchi, East Siberian, Laptev, and Kara Seas and the waters of the Canadian archipelago, including Hudson Bay. The Arctic Ocean so defined is wholly ice-covered for at least six months of the year, but the ice generally clears from the coastal waters of the outer perimeter for a few months in the summer (with the exception of the waters off the Canadian archipelago and northern Greenland, which never clear – see Map 4). The average thickness of the sea ice, taking into account the frequent ridging and hummocking, varies between 3 and 7 metres, depending on the locality, but thicknesses up to 20 m are frequently found. Icebergs, which have a much greater draught than sea ice, are rare in the Arctic Ocean.

The area of the ocean is about 13 million sq km, or five times that of the Mediterranean. The central part is divided into two major basins by the Lomonosov ridge, a submarine mountain range extending from northern Ellesmere Island to the Novosibirskiye Ostrova, and there are depths of over 4,000 m in each basin. But the continental shelf extends up to 550 km from the coast of the mainland of Siberia – unsurpassed anywhere in the world – creating a broad belt of shallow water offshore (see Map 22).

Three water masses are distinguished. A surface layer, 150–250 m deep, is cold (at or near freezing point) and relatively fresh, due to melting of ice and inflow of river water. Below it, down to about 900 m, is a warmer, more saline layer which has been identified as coming from the Atlantic and is called the Atlantic layer. Finally, below 900 m is 'bottom water', which is colder than the surface layer, but does not freeze because salinity is high. Surface circulation is controlled by both wind patterns and density differences, and has already been described (see p. 13).

Relatively little is known about the Arctic Ocean, because of the difficulty of access by observers. Besides the physical difficulty, there is a political one; for while there is clear need for co-operative study and planning of research by scientists of several nations, such a development has been hindered by defence considerations. Most of the information now available has been collected since the Second World War either from scientific camps established on the drifting ice for periods of up to several years, or from aircraft which land at selected points for brief periods. The USSR has been the leader in this work, having organized a total of twenty-three drifting stations between 1937 and 1975, and a very large number of aircraft landings.

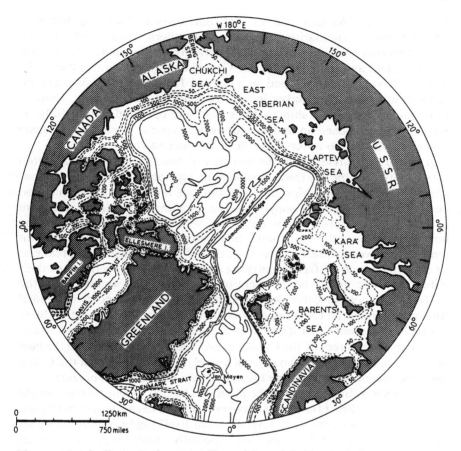

Map 22. Arctic Ocean bathymetry. From *Map of the Arctic region*, 1:5,000,000, American Geographical Society, 1975

Shipping and air routes

The two main shipping routes within the Arctic Ocean are the northern sea route, or northeast passage, along the Asiatic shore; and the northwest passage, along the American shore. These are discussed elsewhere (pp. 58–63, 79–80, and 106–10). Here it is relevant to add some remarks on the possibility of establishing other shipping routes in the Arctic Ocean.

The possibility of navigating in the central part of the Arctic Ocean, either directly across it, or around the outside of the Canadian archipelago and Greenland, is becoming real. The idea of operating submarine freighters here has been mooted since the advent of nuclear-powered submarines, and in particular since the under-ice voyages of USS *Nautilus* and

Skate across the Arctic Ocean in 1958. Shallow water for long distances offshore inhibits submarine operation off northern Alaska and off most of the Siberian north coast. The best operating areas would be the central Arctic Ocean and the waters of the Canadian archipelago. Military submarines have been developed or modified for under-ice work, but no submarine freighter or tanker has got beyond the design stage. There is little doubt that such a ship could be built and operated successfully. A US firm has proposed an underwater vessel capable of carrying 190,000 tonnes of oil, but no orders have been announced yet.

A newer idea arises from the voyages of the supertanker *Manhattan* (115,000 dwt) in the northwest passage in 1969 and 1970. These demonstrated that really big ships can overcome ice more effectively than had been supposed, and lent support to the view that still bigger and more powerful ships, designed for ice work, would be able to navigate the northwest passage the year round. Such ships would have a displacement ten or even twenty times greater than that of today's biggest icebreakers. They would combine freight-carrying and icebreaking functions, so their potential, while unproved, is likely to be great. They would reduce the usefulness of normal icebreakers, whose help would tend not to be needed (though still required for support to smaller vessels). Submarine freighters would likewise not need icebreaker assistance. Since the central Arctic Ocean and the waters of the Canadian archipelago would also be the best operating areas for these two sorts of vessel, it would seem that the USSR has little incentive to examine these possible future developments.

The uses to which they could be put might be two. A shorter route might be required for direct freighting between, say, Western Europe and Japan: or a mineral discovery in the Arctic basin itself might require an outlet of this kind. The first does not look especially imminent at the moment, particularly since the oil companies, which organized the *Manhattan*'s experimental voyages, shelved the idea of carrying oil through the northwest passage. But the situation would change quickly with the discovery of oil in commercial quantities in the Canadian archipelago. It is clear that the economic incentive would have to be great to justify the development costs involved. But once these costs were met and an appropriate class of vessel was available, there would probably be no lack of users.

Air routes, however, have existed for over twenty years. That from Western Europe to Japan via Alaska follows the outer edge of the Canadian archipelago and may, in its northern variant, cross a long stretch of the Arctic Ocean. It is used by half a dozen major airlines. There are no scheduled stopping places in the Arctic on these flights, but some emergency landing fields are available.

Fisheries and hunting

The presence of pack ice is an effective barrier to commercial fishing operations. This may not remain so for much longer, since the USSR is planning to build icebreaking trawlers. But until such ships come into service, the only possible fishing grounds in the Arctic Ocean are in its peripheral waters which clear seasonally, such as the Beaufort Sea, the Chukchi Sea, and the Kara Sea, and these are not significantly exploited. The United Nations Food and Agriculture Organization's *Yearbook of Fishery Statistics* lists only the USSR as a country fishing the Arctic Ocean, and the highest catch there in recent years was 7,400 tonnes in 1968. The problem of icing of trawler hulls and superstructure can be acute in northern waters, and it has not been satisfactorily solved.

Sealing and walrus hunting is not subject to the same limitations, in so far as these animals are generally hunted when they are lying on the ice. But the major populations of commercially exploited seals are not in the Arctic Ocean, so hunting in that ocean is restricted to subsistence hunting by native peoples at a number of points off both the Asiatic and American shores, and to a small commercial operation in the Kara Sea. Walrus are protected, and hunting is restricted to natives. Whaling is in a different category, since it is practised in the open sea. There was at one time commercial whaling along the Alaskan coast, extending into Canadian waters, and in Hudson Bay, but it had virtually disappeared by 1914. The white whale (*Delphinapterus leucas*) was also hunted in the Kara Sea, but this industry, never large, has probably disappeared too. Table 37 shows the extent of subsistence whaling by native groups in North American Arctic waters.

Polar bears wander for great distances across the ice of the Arctic Ocean, and have been hunted by white trophy-hunters as well as by natives. The polar bear is now protected in many countries owing to pressure by environmental groups, since the total number of polar bears in the world probably does not exceed 10,000. The status of the polar bear is under continuing review by a group of biologists from the five countries, who formed a Polar Bear Group within the International Union for Conservation of Nature and Natural Resources in 1965.

Position in international law

On the face of it, there might appear to be no special difficulty about the position of the Arctic Ocean in international law. The regime for the high seas should apply here, as in any other ocean, with uncertainty only as to the width of territorial waters claimed by different countries; and this

Table 37. Arctic whales. Subsistence catch by species and area, 1955–75
(number of animals)

Area and species	1955	1960	1965	1970	1971	1972	1973	1974	1975
Hudson and James Bays									
White (beluga)	484	385	354	326	13	—	178	255	98
Narwhal	—	—	—	—	5	—	11	10	—
Western Arctic									
White (beluga)	n.r.	n.r.	285	n.r.	n.r.	120	297	118	149
Narwhal	—	—	—	20–30	—	31	—	—	—
Total Canada									
White (beluga)	—	—	639	—	—	120	475	373	247
Narwhal	—	—	—	20–30	—	31	11	—	—
USA–Alaska (Kotzebue, Pt Hope, Wainwright, Barrow)									
Bowhead	16	19	6	23	20	39	37	21	
White (beluga)[a]	185	185	185	185	185	185	185	185	185

[a] Annual take not reported. Estimated in *International Whaling Statistics*, vol. LXXI, 1972, p. 27 as Kotzebue 100, Pt Hope 50, Wainwright 25, Barrow 10. Annual fluctuation estimated to range from 80 to 270.
n.r. = not reported.
(*Source: International Whaling Statistics*, Oslo, Committee for Whaling Statistics.)

does indeed appear to be the position in regard to the northern sea route and the northwest passage.

But the presence of a cover of floating ice complicates the matter. There have been attempts to assert sovereignty over the ice, notably by individual jurists in the USSR and Canada. These attempts have not been accorded recognition, even by the governments of the countries concerned. But it is now becoming clear that situations might arise which the body of international law relating to the high seas would be inadequate to deal with. Anyone can land on floating ice, and put it to whatever use he wishes, so long as it remains outside territorial waters. As already mentioned, a number of scientific stations have been established on it. One such station, which had floated round the Arctic Ocean for many years, was ice island T-3 manned by a US party, and it was the scene of a murder in 1970. The case was tried as if the crime had taken place in an American ship. But there are many ways in which the parallel to a ship would break down. The 'crew' might not be a unit at all, and might, for instance, be spread out over many hundreds of square kilometres.

There would seem to be a need, therefore, for a special multinational agreement. If the regime established by such an agreement related specifically to the presence of ice, it need not set precedents which might be embarrassing in other parts of the world. The jurisdictional aspect might be one concern of such an agreement. Others might be such matters as the control of ship design and equipment, navigation instruments, officers' qualifications, traffic regulations, communications, and scientific research related to any of these. Particular research topics which come to mind as suitable subjects for joint study are ice islands, seabed topography, and trans-Arctic aviation as it affects the surface below. In fact, the level of concern shown today for factors causing pollution of the environment brings these other matters into prominence, especially since oil is a possible freight. The Canadian government passed legislation in 1970, with supporting regulations, to control pollution between longitudes 60° W and 141° W, and north of latitude 60° N. This was a unilateral action, but Canada expressed the hope that others would formulate similar controls.

Another important issue may be conservation of wild life in the region. Walrus and polar bear have already been mentioned. Regulatory measures in cases of this kind are likely to be established through separate agreements of a more specialist sort.

CHAPTER EIGHT

The circumpolar north in world affairs

Introduction

'A glance at the map of the northern hemisphere shows that the Arctic Ocean is in effect a huge Mediterranean,' wrote Vilhjalmur Stefansson in 1922. The concept was somewhat startling at that time, for it was by no means clear how a frozen ocean could bring together, rather than divide, the lands round its shores. But time has removed the obscurity, and has fully justified Stefansson's forecast that those in a hurry to go from England to Japan 'will fly over the north polar ocean'. This wedding of geography and technology gives rise to two questions. To what extent has the relative ease of contact between the countries round the North Pole affected political and economic relations, both between those countries and with the rest of the world? And to what extent do they contribute to a sense of community in the north?

The circumpolar north was described in the introductory chapter of this book as an ocean surrounded almost completely by a ring of land, the only substantial break being between Greenland and Scandinavia and the other breaks being the relatively minor straits between the islands of the Canadian archipelago and the 90-km gap between Asia and America, the Bering Strait. The catalogue of physical and geophysical features, climate, and flora and fauna can be similarly generalized as rings or bands around the northern sea, although with more frequent breaks and greater irregularities. Permafrost, tundra, and taiga zones are distributed throughout the land masses on all sides of the ocean, and sea ice is a major characteristic of the ocean itself. Contemporary descendants of mankind's pre-sixteenth century migrations are distributed in roughly similar bands, the Eskimos, for example, ranging from eastern Siberia across Alaska and Canada and into Greenland. Modern western man entered the northern lands and seas in two major movements, northward and westward across the North Atlantic and then across the North American continent, and northward and eastward from Russia across the Eurasian continent, both movements meeting in the North Pacific and Alaska.

In the context of this physical environment and historical evolution, it might be expected that the polar Mediterranean would exert a strong unifying influence on human occupancy and use of the northern lands and seas. When the social, political, and economic dimensions of the circumpolar north are studied, however, the generalized view is one of fragmentation and division of the land into distinct and not always mutually amicable political units. The seas are also politically divided by treaties and conventions into territorial and international areas and a number of intermediate zones. The political divisions not only are extensions of separate national systems with their bases farther south (excepting, of course, Iceland), but they are of unequal political status. Included in these northern neighbours are the Republic of Iceland, the 'sovereign' State of Alaska, and an array of Canadian, Soviet, and Scandinavian territories, districts, and other political and administrative units. At this level it is difficult to discover any polar community.

Before we can speculate on the future of the north in world affairs, it is necessary to review the nature of political relations (boundaries and the forces of sovereignty they reflect) and the nature and orientation of the economic systems at work in the north. While political and economic forces appear on balance to be divisive, there are other forces within the north which tend towards some sense of community. Although in one sense boundaries reflect equilibrium lines between political and economic powers, in another sense they are interfaces across which economic intercourse between people of adjacent political–social entities can take place and between which co-operative arrangements can be made. Growing awareness of shared space and resources in the north is increasing the degree to which conflict in relations is being replaced by co-operation and legally created order. Finally, the very massiveness of the contemporary invasion of economic exploitation from the south, regardless of political origin, has created a sense of shared fate among the established northern residents, native and non-native. In Alaska, for example, the threat posed by the pending petroleum industry invasions brought together for the first time the Eskimo, Aleut, Tlingit, Haida, Tsimshian, and Athapaskan people into a united front, which now appears to be expanding across political borders in response to similar economic threats and resulting in a newly discovered sense of community among the northern indigenous peoples across the northern hemisphere. In November 1973 the first Arctic Peoples Conference was held in Copenhagen and was attended by Eskimos and Indians from North America, Greenlanders, and Lapps. Their discussions centred on land claims, participation in resource development, and the need for recognition of their identity. In 1977 the Eskimo peoples met at the First Inuit Circumpolar Conference, held in Barrow. There were delegations from Alaska, Canada, and Greenland. The conference discussed many political issues affecting Eskimos, and resolved to set up a continuing organization.

Political relations

The Soviet Union has had the fewest political relations and links with other northern countries. Her Asiatic frontier in the north, which runs through Bering Strait, is completely closed, and there is no intercourse between Chukotka and Alaska. Even the Eskimos, who have in some cases close family ties across the strait, are no longer permitted to visit each other. Her European frontier with Scandinavia is on land, and is also an effective barrier. There has been some limited co-operation with Norway and Finland, chiefly in respect of hydroelectric installations, and a small movement of tourists, effectively from Scandinavian countries only, has been permitted.

Greenland has not had close links with her North American neighbours; in fact, there has been almost no contact at all. This is simply explained by the island's constitutional position as part of Denmark, causing Greenland to face the other way across the Atlantic, and by the nature of the Greenland economy, which has had little to offer Canada. This is now changing, however, in that the native people on both sides of Davis Strait are forging closer links, the new lead–zinc mine at Marmorilik in northwest Greenland is controlled by a Canadian firm, and there has been talk, too, of a scheduled air service across Baffin Bay. There are strong links in the Scandinavian north and between Alaska and the Yukon Territory owing to the close relations between the nations of which they form part, few language difficulties, land frontiers, and native groups that straddle them. There is a negative aspect in the case of the Canada–United States link, however, in that the very high level of United States investment in the Canadian north (as in Canada generally, but the north seems in some ways more vulnerable) exacerbates Canadian sensitivity about sovereignty.

The Canada–Alaska boundary presents an interesting case in which the very nature of the border itself promotes the need for closer political and economic arrangements. For 640 miles (1,000 km) it is the 141st meridian, cutting across major mountain systems and river basins, and for the remaining 800 miles (1,300 km) a saw-toothed zig-zag from one reference point to the next. When the Anglo-Russian treaty of 1825 and the Alaska Boundary Settlement of 1903 were negotiated and arbitrated, this allocation without reference to natural features seemed adequate. It did resolve immediate territorial conflicts and in 1825 and 1903 the adjacent lands were uninhabited, unknown, and devoid of recognized economic value. The very arbitrariness and artificiality of the boundary, however, now should bring Alaska and Canada closer together in a range of development and management matters. Precedent for this can be found in the negotiations for the Columbia River Treaty of 1961, which through co-operative development realized economies not possible by independent action, while minimizing adverse consequences for the citizens of both countries.

More generally, however, for two decades after the Second World War the dominating factor in political relations between the circumpolar countries was the fear of hostilities. North America confronted the Soviet Union across the Arctic Ocean. This had two results. One was that northern links between countries in the western hemisphere became closer; between the United States and Canada, principally, but also between the United States and Greenland, and to a lesser extent, between all three countries and Norway, as fellow members of NATO. Svalbard, however, remained apart, because, although Norway exercised sovereignty over it, the provisions of the 1920 treaty forbade any militarization. The other result, expectably, was that there was very little co-operation between any of these countries and the Soviet Union.

The most notable practical effects of this were the construction of many military installations. There were four chains of north-facing radar stations protecting the population centres of North America: the Pinetree line, the Mid-Canada line, the Distant Early Warning (DEW) line, and the Ballistic Missile Early Warning System (BMEWS). The stations were mostly in Canada, but the DEW line extended into Alaska and Greenland, and BMEWS has only three stations, which are in Alaska, Greenland, and Britain. There were also a number of military airfields, in Canada, Alaska, and Greenland. The large United States Air Force base at Thule in northwest Greenland played a key role in the 1950s and early 1960s. There were dozens of smaller airstrips, often without much elaborate equipment, whose function was to permit resupply of the radar stations and other inhabited points. There was a sophisticated communications system, requiring more ground stations. The infrastructure underpinning this military effort was extensive, and involved all three sovereign territories.

The link with northern Norway, though no doubt apparent in strategic thinking, was not manifest in any activities or constructions on the ground, and did not lead to any closer contact between Norway and North America than already existed.

It should be added that although these strategic pressures led to close co-operation between the three countries represented in the North American Arctic, and to contact in the territories concerned, it led also to tensions between them. Canadian interest in the north grew greatly after the Second World War, and the stimulus for this, it must be confessed, was not only fear of what the Russians might do, but concern at what the Americans were already doing. Joint United States–Canadian defence schemes called for the construction on Canadian territory of installations manned and operated by Americans – another cause for Canadian anxiety. A number of subsequent programs of activity in the north undertaken by Ottawa were, in part at least, a reaction to this situation. The Danish government, equally, felt irritation from time to time at the impact of the United States forces in Greenland. These animosities were essentially minor, arising from the actions of individuals assuming unwarranted authority and corrected when known at a higher level.

Nothing has been said, because little is publicly known, about the military posture of the Soviet Union in the north. It seems likely that chains of radar stations exist there too, and there are certainly many military airfields. The whole coastal area is believed to be a military zone, and a major testing ground for nuclear weapons is located on the north island of Novaya Zemlya. None of these facilities has been shared with any other northern country and their political effect has been restricted to the military role they play.

Both the North Atlantic and the North Pacific are major fishing areas of the world, where fishing fleets meet from many countries, including all the circumpolar countries. The actions of the fishing fleets may seriously affect relations between them. The 'cod wars' between Iceland and Britain, fought out in the 1970s in the waters round Iceland, are an unfortunate example. There have been infringements of territorial waters also in the Gulf of Alaska and in the Bering Sea, where Soviet, Japanese, Korean, and United States fishing fleets operate, but in general fishing proceeds without very much quarrelling. Although there will tend to be more as pressure on the resource grows, the establishment of exclusive economic zones should in the long run improve the situation. It may be said therefore that a common interest in harvesting the resources of northern waters has led, slowly but steadily, to a certain measure of practical and political co-operation.

Air transport provides another, less publicized, example of international co-operation of a rather similar kind. Many airlines operate flights across the Arctic Ocean. These flights are controlled from a series of ground stations. The International Civil Aviation Organization allocates authority to a state for regulating certain activities within a given area. Thus Canada or Denmark might be responsible for the safety of aircraft flying far out over the Arctic Ocean, and this responsibility would be passed from one to the other as the flight continued.

The possibility that such co-operation may spread over into other fields of activity is quite a real one. There are no overlapping territorial claims to complicate the matter. The only contentious issue of this kind is the question of sovereignty over floating ice. Some equate floating ice to the sea beneath (that is, it may be either high seas or territorial sea), while others think that the wider range of uses made possible by its solidity ought to confer a status different from that of liquid water. No one has gone so far as to equate it with land, however, and the problem is, for the moment at least, a relatively academic one. Wherever joint action by two or more northern countries can be seen by each to be beneficial, further agreements such as those for fishing and air transport may be expected. The next might be, for instance, the prevention of pollution in the northern environment.

Economic relations and exchanges

Each of the regions within the circumpolar north, with the possible exception of Iceland, exhibits the characteristics of a colonial economic system. The process by which the economy of each was raised beyond resident subsistence to the present level of development required the import of capital and labour, and consequently put the control of the nature and direction of the process in the hands of the sources providing these essentials. Until after the Second World War, political factors dictated that the nations exercising sovereignty over these divisions of the circumpolar north were to be the exclusive beneficiaries of development in each case. This even extended to the provision of independent marine transport systems for Alaska and the western Canadian north, where joint systems would have been of mutual benefit, but were prohibited and still are by United States legislation for the economic benefit of the US merchant marine and shipbuilders. There were variations on and modifications to this basic colonial theme of economic development primarily for non-resident benefit. As has been discussed in the regional chapters, in step with the movements to foster greater political self-determination and self-government for the resident people, the purely colonial economic objectives (securing selected raw materials at lowest cost for the benefit of residents of the metropolitan area) were modified to take increasing account of the economic well-being of local residents as well.

Between the Klondike rold rush and the Second World War there were virtually no direct economic linkages or exchanges between the major northern regions, economic inter-relations being via the mother-country. The joint commercial venture of the North Pacific fur seal fishery under the 1911 treaty between the United States, Great Britain (Canada), Russia, and Japan was one isolated exception. As has been noted, the Second World War allied all the nations represented in the circumpolar north against the Axis powers and Japan, and the post-war period brought a confrontation between the USSR and the North American partners of this former alliance. These defence arrangements have been accompanied by exchanges and movements of goods and personnel, but the motivating force has been military rather than economic.

Starting in the 1950s and accelerating in the decades following, a new kind of colonialism, on an international or multi-national scale, has begun to emerge in the non-Soviet north. An uncritical examination of such foreign trade statistics as exist for the circumpolar regions (for instance the Alaska import–export data summarized in Table 23) might suggest that there is a trend towards freedom of the former colonial areas from their 'mother countries' as evidenced by the increased portion of the gross product generated in each region which goes directly into foreign as opposed to domestic trade. Deeper examination, however, indicates that the 'metropolitan country' has merely been expanded. In Alaska, for example, it was determined in the 1950s that permitting (or even encouraging) Japanese investment in the development of Alaska's forest resources was in the national interest in fostering closer political ties with that nation. In the case of the Canadian north, the presence of foreign investment in petroleum and mineral developments is a reflection of domestic capital shortages, but here the people in both the north and south of Canada are clearly aware of the spectre of international colonialism.

The exploitation of northern energy resources adds a further dimension to the new northern colonialism. The survival of Western industrial society as it is now organized and fuelled is at stake. Until technological revisions can be made, the continued supply of an increasing flow of oil and natural gas for energy purposes is of critical importance to this survival. Investment costs are staggering and prospects of returns correspondingly high. The scale of need, cost, and profit assume international proportions and it is not surprising that in the Alaskan petroleum development British Petroleum is one of the major partners, that the decision was made not to build an all-land transport system which would have provided access only to domestic United States markets, and that Japan, the source of the hundreds of kilometres of pipe and the pumping and docking facilities, should be considered as recipient of a part of the final product.

The future of real economic exchanges directly between the regions of the circumpolar north will be a function not so much of their geographical contiguity, but rather of the size of the economic markets in each region, the degree to which the natural resources concerned are processed, and political considerations. The economic exchanges that take place between the north and south will continue to overshadow such lateral connections.

Scientific co-operation

The world draws together at the Poles, and the circumpolar countries share much the same geographical conditions. They are cold, the days are long in the summer and short in winter, precipitation is light, the sea is frozen except for a few weeks, and in most of the area the ground below the surface remains frozen throughout the year. This similarity in geographical conditions is reflected in the biology, and many species of fauna and flora are circumpolar in distribution. Man has entered the north and evolved successful material and social cultures to meet its special conditions, cultures that have much in common. Some geophysical phenomena are centred on the poles, observations of others throughout the north are necessary to complete worldwide coverage.

The north has other, less obvious, attractions for scientists. The sparseness of the population and the absence of many of the structures of social life which exist in well-settled regions farther south make it relatively easy to identify and study the economic and social behaviour of particular groups of persons. It is much easier to observe, say, the reaction of a minority people, be they Copper Eskimos, Skolt Lapps, or reindeer-herding Chukchi, to a new employment option, if there are very few other persons or employers in the area. The ecological systems of lakes become simpler and therefore easier to study as the number of species they contain becomes less. The effects of physical constraints, such as cold, are more readily observed where the constraints are most extreme. In many ways like these Arctic studies have some particular importance and there is a need for international co-operation to provide comparisons, to give better coverage, to exchange ideas, and to share the scientific load.

International co-operation in polar science got a good and early start in the discussions that led to the International Polar Year of 1882–3, during which eleven countries operated twelve stations in the Arctic to carry out a variety of geophysical observations. Attempts during the next fifty years to form a permanent international polar scientific organization failed, but the Second International Polar Year of 1932–3 revived interest in many countries. The value of this co-operative effort was attested by the decision, taken shortly after the Second World War, that the next polar year should be held twenty-five rather than fifty years after the previous one, and that it should be widened to cover the whole world. The International Geophysical Year of 1957–8, in which sixty-seven nations worked together, grew out of the International Polar Years. After the International Geophysical Year ended, the International Council of Scientific Unions established a Scientific Committee on Antarctic Research to continue the scientific co-operation in the Antarctic carried out at that time.

International science has been moving forward on other fronts, often with UNESCO encouragement, through continuing scientific agencies such as the World Meteorological Organization and the International Union of Geodesy and Geophysics, and by special limited programs such as the International Hydrological Decade and the International Biological Program. Sometimes it has been a matter of co-operation between a few nations which share a common concern. Examples are the decrease in fur seals in the North Pacific leading to the scientific management of the resource by the United States, and the Titanic disaster which resulted in the International Ice Patrol conducted by the US Coast Guard on behalf of the nations concerned.

No organization has emerged for the Arctic corresponding to the Scientific Committee on Antarctic Research. Both the geographical and political situations are very different. The north, lying between the major powers, has great strategic significance. It is logical therefore to expect a number of defence installations, together with security measures in their immediate vicinity. In the case of the USSR, however, the whole of the north is denied to visitors from the west, and many categories of scientific information are withheld. This has been and remains the great stumbling block to full international scientific collaboration in the north.

The USSR has participated in several international research programs in the north such as the International Geophysical Year, and took the initiative in organizing 'Polex', an investigation of thermal and dynamic processes in the polar regions. She has also entered into agreements with foreign countries for the exchange of Arctic information and scientists, but these are usually bilateral, and she has appeared reluctant to implement them except in fields where it is clearly to her advantage. As a result little is known for instance on how her native people are adjusting to the rapid changes they face, an area in which other countries could have much to learn from Soviet experience.

Apart from in the USSR there are few obstacles in the way of scientific co-operation in the north, and throughout the rest of the area scientists from other countries are usually welcomed, whether they want to run their own expeditions or to participate in a joint endeavour. This carries with it concern that the host country may be left without some of the scientific knowledge and expertise she requires to operate effectively in her own north. This was one of the considerations that in 1976 induced the Canadian government to issue guidelines for scientific activities in northern Canada.

These Canadian guidelines draw attention to the need for Canada to play a significant role in international Arctic research in view of her own large Arctic territories. While Canada is naturally most concerned with research that is directed towards Canadian objectives, she recognizes an obligation to assist the international scientific community in projects that have little Canadian priority. She has therefore adopted the following policy.

Where the Federal Government initiates international co-operative scientific activities in the Canadian north, the following principles should apply:

(a) the Canadian contribution should be defined in terms of Canadian objectives;
(b) the leadership in co-ordinating such activities in Canada and their effective control should be provided by Canada;
(c) Canada should receive all data and all analytical results.

Where the initiative for co-operative international programs comes from other countries and the objectives are not priority items for Canada, the following principles should apply:

(a) government logistic support of international scientific programs should not be considered a substitute for scientific involvement;
(b) the need for the program and the reason for conducting it in Canada should be stated to the satisfaction of Canadian authorities;
(c) there should be Canadian scientific participation in any significant scientific investigation in the Canadian north;
(d) non-government sources, primarily universities and scientific institutions, should be invited to participate;
(e) Canada should receive all data and all analytical results.

These guidelines should be useful in preventing misunderstandings and in removing what could become a source of difficulty in international co-operation, especially when nations of different sizes and scientific capabilities are involved. They open northern Canada to foreign scientists, but define the conditions they are expected to observe. Though the principles were designed with major scientific programs in mind, they can be applied equally well to small foreign scientific parties. In most respects the guidelines are a statement of what has become the practice in other northern countries except the USSR.

Few scientific investigations are conceived in the first instance by governments. Most have their origins either with individuals or in meetings between people with related interests. In most northern countries scientific organizations have evolved, which bring together those who share an interest in the Arctic regions. North America has many such groups, usually centred on a university. Some have government backing. All countries with Arctic territory have such institutions, and so do some with none (Britain, France, Germany, Italy, Japan). Although all these centres are rooted in their own country, and only one is even bi-national, most have informal contacts which bring them into close and regular touch with interested persons in other countries. In this unplanned way effective forms of international scientific co-operation have frequently evolved.

The northerners

Although the distinction between 'natives' and 'whites', or between 'aboriginals' and 'immigrants', seems at first quite straightforward in many parts of the north, problems arise as one examines the question more closely. Archaeological discoveries yield evidence of earlier and unsuspected aboriginals, while the immigrants have arrived at widely spaced intervals, and the aboriginals themselves were immigrants at one time. Physically, the distinction between native and white is relatively marked in North America, but is much less so in Scandinavia and European USSR, where peoples like the Lapps and the Komi are not only white, but not at all easily distinguished from their neighbours the Finns and the north Russians. One definition is that immigrants are those that arrive in the north from an on-going developed southern culture and who continue to draw support from that culture.

Applying this distinction, one can say that the native population of the north is now about 1 million, and that there are seven to eight times that number of immigrants. Yet forty years ago the natives, who were almost as numerous then as now, were a majority in all regions except Scandinavia (where the line between the two groups is the least meaningful). The figures are given in Appendix 2.

The increased number of immigrants into the north has reinforced their contacts with the south. Northern whites now have much the same interests as their countrymen in the south and have lost some of the long-term commitment to the north that had been characteristic. Many have in fact arrived in the north to seek their fortunes on a 'get rich, get out' basis. Defence-related populations are typically residents for limited tours of duty, and petroleum-development-related populations are members of multi-national industrial groups and subject to movement to any part of the globe in response to international exploration, development, and production planning. They know little about the north and have little understanding of the native people. For them progress and development are synonymous.

If there is still a northern individuality, it is specific to each country. Despite the similarity of the conditions under which they live and of their roles, the immigrants in one northern country show less interest in their counterparts in other circumpolar nations than they did twenty-five years ago. Then, for example, Alaska and the Yukon appeared to be growing closer to one another, and aviation was then seen as a cohesive force within the north rather than as a way of bringing in southern influences, ever more pervasively.

The northern natives have been brought into much closer contact with a civilization that is totally alien and yet exercises attraction because it appears to hold promise of material rewards. Often it is forced upon them by circumstances they cannot control. Development is frequently justified on the grounds that it gives the natives the choice of whether to continue their traditional life or to enter the bright new world of wage employment. It is, however, rarely a fair choice, for the cards are stacked against the traditional life, which receives little or no support compared with the massive incentives provided to industry. The choice offered to the natives becomes simply whether or not to survive.

Improved communications have also shown the native population the extent to which their ways differ from the southern ways. Many have visited the south and are familiar with the failings of western civilization, when previously they had learned mainly of its successes. In most cases native populations have become small minorities in their own lands, and they have begun to seek strength, encouragement, and support from one another.

The situation is now dynamic. There is a series of problems, whose form and severity vary in different parts of the north, and many attempted solutions. In North America, the 'native peoples question' has become probably the most important single issue. Although the natives are no longer in the majority in Alaska and the white population in the Canadian northern territories has been increasing rapidly, the proportionate drop in native population is more than offset by the acquisition of a political voice. The process is farthest advanced in Alaska, where the natives, though now constituting only about 18 per cent of the population of the State, have made significant gains through native organizations, particularly the Alaska Federation of Natives, formed in 1967. The native newspaper *Tundra Times*, launched by a well-wishing and far-sighted American doctor in 1962, played a very important part in defining issues, consolidating and explaining the native position, and working for understanding on both sides. The most recent, and most spectacular, achievement has been the Alaska Native Claims Settlement Act, which became law in 1971. Under this Act, the native peoples may select 40 million acres (160,000 sq km) of land, and will receive over a period of years nearly $1,000 million, most of it to be paid to native corporations. In Alaska, the hope is expressed that it is through political and economic organization that the native peoples will be able to find their own way of coming to terms with white civilization.

In the Northwest Territories and the Yukon Territory of Canada, natives still constitute a larger fraction of the population than they do in Alaska, but political awareness is at a lower level. Native organizations are still nascent, and it is not at all easy to foster unity among the very widespread and disparate settlements, or to exercise leadership in a society where imposing one's will on others provokes dislike and ridicule rather than respect. The Indians, though a minority of the northern natives, can make common cause with the much larger group of Indians in Canada as a whole, and therefore have a stronger position. Both groups are vulnerable to the harmful aspects of culture contact. The Canadian government has been aware of the problems, and its attention has been directed to them once again by the electorate, which has expressed deep concern. But awareness and concern are not yet solutions, and it is fair to say that a policy based on radical rethinking of the whole relationship between whites and natives has not yet started to emerge. Undoubtedly one of the difficulties has been how to find out just what the natives themselves think and want; it was much easier to do things for them, based on a judgment by whites as to what their needs should be. Here, the emergence of a native association which can speak for all, or a large number, of natives, will clearly be a great help. In general, the Canadian government has now adopted the position of welcoming, and encouraging with financial aid, the growth of more effective native involvement; and, indeed, some of the local native associations are now developing fast. A significant step in the direction of ascertaining native views was taken by setting up the Mackenzie Valley Pipeline Inquiry of 1974–7 ('The Berger Commission').

The situation in the Soviet Union is, from the point of view of pro-portionate numbers, not unlike that in Alaska. The natives, who were over half the population in 1926, now form less than one-sixth. The absolute numbers involved, however, and the size of the territory, are much greater than in the other sectors: almost 90 per cent of all northern peoples in the world live in the Soviet Union. It would be of the greatest interest, therefore, to know how they are getting on. Are they coming to terms with technological civilization with greater or less difficulty than are the Eskimos, Indians, and Greenlanders? Is the conflict between rising material expectations and traditional national values more or less acute there than elsewhere? It is difficult to find out, because on the one hand the Soviet government claims to have eliminated these difficulties, with the result that little evidence of their continuing existence would be likely to reach the press, and on the other hand non-Soviet specialists have rarely been permitted into their homelands. There is, of course, much evidence, in the form of Soviet publications, that demonstrates the success of the transition. Indeed, one would in many cases expect it to be successful. The larger national groups, with a more or less well-developed sense of nationhood and a strong national culture, can probably retain their sense of identity at the same time as they accept new ways and a great many new neighbours. There is no doubt at all that the Komi (322,000 in 1970) and the Yakuts (296,000) have done this successfully. On the other hand the Karelians (146,000), who are the only other group larger than those found in the non-Soviet northlands, are declining in numbers. The real interest centres on the fifteen or so smaller peoples, whose numbers run from 29,000 down to a few hundreds. Their numerical growth between 1959 and 1970 was about the same as that of the Soviet Union as a whole (15 per cent compared to 16 per cent); but northern peoples in North America and Greenland were increasing at over twice that rate. The small absolute size of these Soviet peoples renders them particularly vulnerable; and they also started this period of southern immigration at a lower level of sophistication than the larger groups. More evidence is required before it can be said with assurance that they have surmounted the difficulties faced elsewhere.

The Greenland situation is unlike that in any other part of the north, in that there the natives have remained an easy majority. The distance which separates Greenland from the mother country, and Danish policy in respect to Greenland, are likely to permit this situation to continue. For a long time, the Danish government sought to shield the Greenlanders from harmful outside influences, and to this end exercised strict control over access to the country. Since the constitutional change of 1953, when Greenland became a part of Denmark, this control has been relaxed, and Greenlanders now resent any manifestations of 'paternalism'. The efforts of Denmark over two centuries, though undoubtedly well-meaning, have not led to a completely harmonious situation. But although there are complaints and problems, it is still arguable that the social situation for natives in Greenland is better than in most other parts of the north.

In northern Scandinavia, the Lapps have long been a small minority. There the distinction between aboriginals and immigrants becomes somewhat hazy, for some of the latter have been there a thousand years. It is true that the Lappish language is still quite widely spoken (though not by all who call themselves Lapps), and that there is a pan-Lappish movement interested in asserting Lapp rights. But most Lapps gain their livelihood in just the same ways as most immigrants and the factors that are common to the two groups greatly exceed in importance the factors that distinguish them. Although Lappish problems can be shown to exist, it is a point for argument as to whether they are primarily due to any specifically Lappish cause, or whether they spring from the fact that the locality in which the people live is poor, or depopulated, or in any other way underdeveloped. The situation in Scandinavian Lapland (leaving aside the 2,000 Lapps in Soviet territory) is an example of an aboriginal–immigrant mix which has continued for longer than in any other part of the north. It is the furthest evolved in the series of cases we have been considering; but that does not mean to say that the others are likely to evolve along the same lines.

Iceland is included in Appendix 2 only for the sake of completeness.

Iceland had no aboriginal population when the first Irish and Norse came, and its population today descends from those early immigrants and their later reinforcements from Scandinavia. No subsequent wave of immigrants has been culturally so dissimilar as to push the first wave into the position of aboriginals.

It is interesting to note that a way of limiting the incoming white population is practised in three northern countries. In the USSR, economic considerations are leading to use of the shift system, by which workers live in the south and work shifts in the north. In the North Slope oilfield in Alaska and in oil exploration work in Canada the same system operates, partly at least because that is the way the workers like it. Such limitation of white residents may have the political advantage of reducing harmful impact on native culture – an aspect to which attention was drawn in the Canadian chapter.

The future of the north in world affairs

There are a number of reasons for studying the economic and social aspects of the north. The most obvious is that this can serve as a means of discovering the possibilities for development and of planning the future of the northern regions more effectively. As noted earlier, the north also can serve as a kind of natural laboratory, in which processes may be observed and theories tested. The north can be studied not only for its own sake, and to find solutions to its own problems, but in the hope of throwing light on problems of wider application also. In so far as these wider problems are elucidated, the impact of the north is correspondingly enlarged. Finally, in the introduction to this book some of the principal ways in which the northlands seem likely to make an impact on the rest of the world were suggested and it is to an examination of the future of these impacts that attention is now turned.

Three ways in which the north will in varying degrees bear upon the rest of the world were discussed. Of great importance is the provision of raw materials, in particular energy in oil and natural gas. Somewhat less is its importance as living space, not so much in the conventional form of settlement, but rather as havens of essential solitude, natural beauty, and absence of pollution and congestion. Finally, its location will give the north an importance in sea as well as air transport.

The significance of the natural resources of the north can be expected to increase as world demand rises and as technology overcomes the environmental problems and costs of northern development. Pressures on resources of raw materials will continue to mount, because for many years at least world population will continue to increase. Those who would like to see the strain on the earth's finite resources diminished can hope only to reduce the rate at which it is increasing. Even that will require an immense effort. To reduce the strain in absolute terms seems quite un-realizable in the foreseeable future, unless a catastrophe of very large dimensions physically annihilates a significant proportion of the world's inhabitants. We must therefore expect more and more demand. It will be primarily for minerals, including petroleum, because natural conditions are generally unfavourable for large-scale exploitation of renewable resources.

In practical terms, the demand will be expressed as increased geological exploration, increased petroleum and mine development, and continuous reassessment of known but unworked deposits in the light of changing prices. This is what has happened over the last twenty-five years. Although there have been fluctuations in the curve, the trend has been steadily upwards. Transport technology has gone a long way towards overcoming the former liability of remoteness and the technological means are known to be, or likely to be, available to meet most of the problems posed by the environment. A decision not to exploit a given deposit is therefore likely to result from economic or social factors. This holds good for all sectors of the north.

The population of the north has not grown significantly in response to programs designed to foster settlement for its own sake, but it has grown as a by-product of strategic and resource development. This is expected to continue to increase at varying rates in the several regions of the north. In the USSR, expansion is officially foreseen at a greatly accelerated rate. Similarly, in Alaska, population is expected to expand by a factor of 2.5 between 1975 and 1995 based on probable oil and gas developments alone.

In the Canadian north there is not likely to be an expansion on anything like this scale. In the immediate future, there may be a flattening of growth curves as effect is given to the government's policy of according priority to the problems of the northern natives, rather than to exploitation of mineral resources. This is also likely to be true in Greenland, Iceland, and Scandinavia; but in Iceland and some parts of Scandinavia the population increase is at present very small, and there may even be a fall in absolute terms locally.

Countervailing forces have been emerging, opposing these anticipated trends, and there is a growing debate between those who see economic development as something which is self-evidently socially desirable, and those who question this approach. There are powerful arguments behind each point of view; a growing world population must have more raw materials, yet the bigger it gets, the more necessary is the provision of natural beauty and the chance of solitude. The conservationist ethic, which now commands considerable political power in the West, is a major factor in slowing the rate of development in the North American north. The

USSR has felt the impact only very slightly, and is not likely to be seriously affected by these ideas until her standard of living has risen significantly. Even then, the difficulty which public opinion has in making itself heard may further delay any effect on policy.

In the West, the industry which may be judged to be the most compatible with a conservationist outlook is tourism. The rising economic value of wilderness is largely conditional upon the possibility of people being able to see and enjoy it. Tourism is already well developed in Scandinavia, Iceland, and Alaska. Greenland and Canada are making a beginning, and even in the USSR there are now sea and river cruises into the Arctic. Tourism is by no means an ideal economic base in this sort of context, and unless it is carefully controlled it may lead to the destruction of the resource it seeks to exploit. But there is little doubt that it will grow.

Renewable resources were discussed earlier as unlikely, for natural causes, to be the subject of large-scale exploitation (their role in meeting local demand, however, will remain important). An exception to this may be water. Many great rivers, in America and Eurasia, flow northwards. There has been much discussion in America on the subject of diverting some of the flow southwards, but no action has been taken, partly because there is an international boundary to be crossed. In the USSR proposals have been made to use northern water to stabilize the level of the Caspian Sea and to irrigate Central Asia.

These land-based developments will constitute the major economic impact of the north on the south, but they are likely to be accompanied by corresponding activities at sea. The stage has now been reached at which ships able to navigate in much of the Arctic Ocean all, or nearly all, the year round could almost certainly be built; that is to say, once again the technology is available if anyone is ready to pay for it. It so happens that the first possible users of such ships – the oil companies interested in the north Alaskan oil – decided against the idea. But it is unlikely that the possibility will remain unexploited for long. Other cargoes or other places may provide the necessary impetus at any time.

But these communications would cross the high seas, and this at once brings up issues which are not just economic, but at the level of inter-national politics. The high seas in the polar regions are unlike those elsewhere in that they have a solid crust of ice, which both impedes ships and permits activity on its surface. It is not necessarily a straightforward matter to apply the law of the high seas to such an area. Up to now, there has not been very much activity – some scientific parties have established camps on the ice, some explorers have traversed it, some ships have penetrated the edges – so few problems have arisen. But in 1970 the murder committed at one of the scientific stations immediately caused jurisdictional problems to become apparent. As activities increase, and ships penetrate more deeply, legal difficulties will surely arise. Issues of environmental pollution, especially oil spills, are already causing anxiety and have led Canada to introduce protective legislation. There is a need for an agreed international regime for the Arctic Ocean and its adjacent waters, and whatever happens in the future can only, it would seem, underline this need.

International co-operation is desirable in many other contexts. A sphere in which it has become important is science. Besides the major international programs, there have been many recent examples of co-operation between two or three countries on specific projects, whether expeditions or longer-term associations. The number and scope of such activities can only grow. This is not to say that national research programs are likely to diminish in importance. But the bigger problems are likely to yield more readily to a broad-based, co-ordinated attack. The results should be advances in knowledge of major geophysical and biological processes affecting our planet and the lives we lead on it, and better understanding of this kind is one of the most important ways the north can affect us.

If the internationalist approach has been successful in science, the question will be asked whether it cannot be extended to cover a wider field, as has been done so successfully in the case of the Antarctic Treaty of 1960. The difficulty of securing agreement to a similar treaty in respect of the Arctic is, however, considerable. Antarctica has no native inhabitants, and so far no mineral resources have been found which are yet worth exploiting. It has minimal strategic importance, and although almost all the land was subject to a claim of sovereignty, the twelve countries concerned found it possible to allow their claims to be 'frozen' for a period of thirty years. In the Arctic, the situation is very different. All land is an integral part of a technologically advanced state, with northern populations and exploited resources. Certainly it would be very desirable to obtain a measure of international agreement over what is to go on in the Arctic. But the first step towards this can probably be no more than an agreement about certain measures of control over the high seas of the Arctic Ocean, mentioned earlier. Even this may not be easy. It is only when all interested parties – there might initially be five or six – are convinced that it would be to their own advantage to conclude an agreement, that any progress can be made.

The strategic aspect of the northlands need not be stressed in this short summary. Quite apart from the hope that mankind will not feel obliged to continue pouring great sums of money into installations which rapidly become obsolete, the undeniable reduction in the strategic importance of the area over the last decade or so, owing largely to changes in weapons, shows no sign of being reversed. That importance lasted less than twenty years, but came at a time of economic expansion and therefore funnelled huge resources into the region. The northlands gained indirectly from some of this expenditure, but now that it has been reduced, are not likely to suffer unduly as a result.

It is unlikely that the circumpolar countries will ever become an exclusive group. All of them, perhaps excepting Iceland, have 'southern' interests which are more significant to the nation than their 'northern' interests. They will enter into agreements with other countries on the basis of many areas of shared concern, and northern lands and seas will be only one of these, in most cases rather low on the list. Furthermore, countries outside the group may have valid interests in the north. So the polar Mediterranean does not have a unifying pull, and a 'polar club' or community among the powers is not likely to emerge. Nevertheless, the northern countries share many interests, and it is for their statesmen to ensure that these become the basis for co-operation rather than division.

Appendices

Statistical data on the whole area covered by the book

Appendix 1. Land area and most recent population figures by administrative divisions

	Area (sq km) ('000)	Popula- tion ('000)	Density (persons per sq km)	Main towns. The first-named in each group is the regional centre (population in thousands)	
				USSR 1976	
Murmanskaya Oblast'	144.9	930	6.4	Murmansk 369	Kirovsk 38
				Apatity 56	Olenegorsk 24 (1974)
				Monchegorsk 46	Zapolyarnyy 22
				Kandalaksha 43	Nikel' 21
				Severomorsk 41	Polyarnyy 15
Karel'skaya ASSR	172.4	735	4.3	Petrozavodsk 216	Sortavala 20
				Segezha 29	Belomorsk 17
				Kondopoga 28	Medvezh'yegorsk 17
				Kem' 21	
Arkhangel'skaya Oblast' (excl. Nenetskiy NO)	410.7	1,407	3.4	Arkhangel'sk 383	Koryazhma 33
				Severodvinsk 180	Onega 25
				Kotlas 62 (with satellites, 110)	Nyandoma 23
				Pervomayskiy 34	Vel'sk 22
Nenetskiy NO	176.7	41	0.2	Nar'yan-Mar 18	
Komi ASSR	415.9	1,053	2.5	Syktyvkar 157	Inta 51
				Vorkuta 96 (with satellites, 200)	Pechora 39
					Sosnogorsk 27
				Ukhta 78 (with satellites, 110)	Komsomol'skiy 17
					Severnyy 15
Yamalo-Nenetskiy NO	750.3	126	0.2	Salekhard 26	
Khanty-Mansiyskiy NO	523.1	425	0.8	Khanty-Mansiysk 26	Nizhnevartovsk 63
				Surgut 67	Nefteyugansk 31 (1973)
Taymyrskiy NO	862.1	43	0.05	Dudinka 23	
City of Noril'sk		168			
Evenkiyskiy NO	767.6	14	0.02	Tura 5	
Yakutskaya ASSR	3,103.2	779	0.3	Yakutsk 143	Aldan 18
				Mirnyy 26 (1973)	Lensk 17
Magadanskaya Oblast' (excl. Chukotskiy NO)	461.4	308	0.7	Magadan 112	
Chukotskiy NO	737	125	0.2	Anadyr' 10	M. Schmidta 3 (1977)
Kamchatskaya Oblast' (excl. Koryakskiy NO)	170.8	323	1.8	Petropavlovsk-na- Kamchatke 202	
Koryakskiy NO	301.5	34	0.1	Palana 4	
	8,997.6	6,511	0.7	*Note*: towns in USSR of under 50,000 are 1970 figures unless otherwise specified.	
Statistical handbook's 'Far north' (1974)	10,900	6,200	0.6		
Slavin's 'Soviet north' (1970)	11,100	6,700	0.6		

(Both the above include fractions of other administrative regions. See notes, pp. 287–8)

Appendix 1 continued

	Area (sq km) ('000)	Popula-tion ('000)	Density (persons per sq km)	Main towns. The first-named in each group is the regional centre (population in thousands)	
				Canada 1976	
Northwest Territories	3,379.6	42	0.01	Yellowknife 8.2 Hay River 3.2 Inuvik 3	Fort Smith 2.3 Frobisher 2.3
Yukon Territory	536	21	0.03	Whitehorse 13	
	3,915.6	63	0.016		
Hamelin's 'Middle north', 'Far north' and 'Extreme north' (1971). See notes, pp. 287-8	7,100	292.7	0.04	Besides the six above: Thompson 17, Flin Flon 9, Labrador City 12, McMurray 15 Happy Valley 8, Gagnon 3.4, Wabush 3.8, Schefferville 3.4 (1976)	
				USA 1976	
State of Alaska	1,519	413	0.3	City and Borough of Juneau 19 Municipality of Anchorage 185 Fairbanks 30 (Fairbanks North Star Borough 52)	Ketchikan 8 (Ketchikan Gateway Borough 11) Kenai 5 (Kenai Peninsula Borough 17) Bethel 3
				Greenland 1973	
Vestgrønland (excl. ice sheet)	119.1	44	0.4	Godthaab 8.3 Holsteinsborg 3.7 Egedesminde 3.4	Jakobshavn 3.2 Julianehaab 2.9 Sukkertoppen 2.9
Nordgrønland (excl. ice sheet)	106.7	0.7	0.007	Thule 0.3	
Østgrønland	115.9	3.1	0.03	Angmagssalik 1	
Ice sheet	1,833.9	—	—		
	2,175.6	49.5	0.02		
				Scandinavia	
				Norway 1974	
Svalbard	62	3.5	0.06	Longyearbyen 0.9	
Finnmark	48.6	79	1.7	Vadsø 5.9 Alta 12	Sør-Varanger (Kirkenes) 11
Troms	26	143	5.7	Tromsø 43 Harstad 21	Lenvik 11
Nordland	38.3	243	6.7	Bodø 31 Rana (Mo) 26 Narvik 20	Vefsn 13 Vestvågøy 11

Appendix 1 continued

	Area (sq km) ('ooo)	Popula- tion ('ooo)	Density (persons per sq km)	Main towns. The first-named in each group is the regional centre (population in thousands)	
			Sweden 1974		
Norrbotten	98.9	262	2.6	Luleå 64	Boden 27
				Piteå 38	Gällivare 26
				Kiruna 31	Kalix 18
				Note: All Swedish entries are for communes, which include surrounding country.	
			Finland 1974		
Lappi	99.1	196	2	Rovaniemi 28	Kemi 28
Total, Scandinavian north	372.9	927	2.5		

(This approximates to the area covered by the Scandinavian term 'Northern Cap' (Nordkalotten).

			Iceland 1973		
	103	213	2	Reykjavík 81	Hafnarfjörður 9.5
				Kópavogur (suburb of Reykjavík) 11	Keflavík 5.5
					Vestmannaeyjar 5.1
				Akureyri 10	

Summary for areas approximating the coverage of this book

	Area	Population	Density	
'Soviet north' (1970)	11,100	6,700	0.6	
'Middle north', 'Far north' and 'Extreme north' of Canada (1971)	7,100	292.7	0.04	
Greenland (1973)	2,175.6	49.5	0.02	
Alaska (1976)	1,519	413	0.3	
Scandinavian north, incl. Svalbard (1974)	373	927	2.5	
Iceland (1970)	103	212	2	
TOTAL	22,370.6	8,594.2	0.38	

Appendix 2   Aboriginal and immigrant population, 1926–31 and 1969–72 (thousands)

Area	Period 1926–31			Period 1969–72		
	Total pop.	Native pop.	%	Total pop.	Native pop.	%
	1926			*1970*		
USSR: Total of fifteen administrative divisions listed in Appendix 1, and of their rough territorial equivalents in 1926	1,469	803[a]	55	5,929	897[a]	15
	1927–31			*1970–71*		
Canada: Northwest Territories and Yukon Territory	Eskimo 4 Indian 6			Eskimo 12 Indian 9		
	11	10	90	53	21[b]	40
	1929			*1970*		
USA: Alaska	59	30	51	303	51[c]	18
			1926			*1972*
Greenland	15	14[d]	95	49	40[d]	82
	1930			*1970–72*		
Scandinavia: Total of six administrative divisions listed in Appendix 1	Norway 19 Sweden 6 Finland 2			Norway 21 Sweden 10 Finland 3		
	650	27[e]	4	913	34[e]	4
	1930			*1970*		
Iceland	109	nearly 100		202	nearly 100	
TOTAL	2,313	884	38	7,446[f]	1,042	14

[a] The assumption is made, which is nearly true, that all members of the northern peoples listed in Table 3 lived in these fifteen divisions.
[b] Another 5,000 Eskimo lived outside these Territories, chiefly in Quebec and Labrador. The Indian total includes only treaty Indians and should be increased by perhaps up to 50 per cent to include non-status Indians (who believe themselves to be as thoroughly Indian as the others).
[c] 28,000 Eskimo, 16,600 Indian, 6,600 Aleut (corrected census data).
[d] Persons born in Greenland, of whom the great majority are Greenlanders.
[e] All Lapps, most of whom live within the six divisions.
[f] The discrepancy between this figure and that given in the summary at the end of Appendix 1 (8,594 thousand) arises largely from differing positions for the southern boundary of 'the north' in the USSR and Canada.

Appendix 3. Fossil fuel production, 1960–75

Area	1950	1960	1965	1970	1971	1972	1973	1974	1975
Oil Tonnes ('000)									
USSR: Far north	—	2,400	5,600	42,000	57,000	75,000	101,000	130,000	161,000
Canada: Northwest Territories and Yukon Territory	25	70	104	130	140	127	128	138	146
USA: Alaska	negl	80	1,590	11,945	11,255	10,508	10,448	10,326	10,283
Oil Barrels per day ('000)									
USSR: Far north	—	48	110	840	1,140	1,500	2,030	2,600	3,200
Canada: Northwest Territories and Yukon Territory	0.5	1.4	2.1	2.4	2.8	2.5	2.6	2.7	2.9
USA: Alaska	negl	1.6	30.5	229	216	201	200	198	197
Gas Cu m ($\times 10^9$)									
USSR: Far north	negl	negl	1.4	18	23	28	36	47	58
Canada: Northwest Territories and Yukon Territory	—	negl	0.002	0.003	0.06	0.4	1.2	1	1.1
USA: Alaska	negl	negl	0.25	3.7	3.6	3.6	3.5	3.6	3.9
Gas Cu ft ($\times 10^9$)									
USSR: Far north	negl	negl	49	630	810	980	1,260	1,640	2,030
Canada: Northwest Territories and Yukon Territory	—	negl	0.6	0.9	2.1	16	41	36	38
USA: Alaska	negl	negl	8.7	131	126	127	125	126	39
Coal Tonnes ('000)									
USSR: Far north	n.a.	28,000	31,000	35,000	34,000	34,000	35,000	35,000	37,000
Canada: Northwest Territories and Yukon Territory	3.7	7.7	8.8	10.9	21	18.4	19.6	17	17.1
USA: Alaska	373	600	804	498	633	606	752	635	695
Norway: Svalbard									
Norwegian (production)	364	404	426	484	456	431	412	461	422
Soviet (shipments)	187	481	393	442	438	446	459	442	456

n.a. = not available
negl = negligible

Appendix 4. Domesticated reindeer, 1940–76 (thousands)

	Year	Number		Year	Number
USSR	1941	1,911	Greenland	1953	0.3
	1951	2,051		1960	2.6
	1961	2,092		1970	3
	1970	2,449		1971	2
	1971	2,401		1972	1.8
	1972	2,356		1973	0.8
	1973	2,345	Norway	1939	130
	1974	2,236		1949	126
	1975	2,196		1959	173
Canada	1940	5		1969	129
	1950	7		1971	157
	1960	6		1972	138
	1965	7		1973	142
	1970	4		1974	133
	1971	5		1975	138
	1972	5	Sweden	1965	188
	1973	5		1970	210
	1974	6	Finland	1939	232
	1975	8		1950	86
	1976	6		1960	181
USA: Alaska	1940	250		1965	193
	1950	25		1970	145
	1969	31		1971	210
	1976	9–14		1972	229
		(estimate)		1973	240
				1974	234
				1975	175

Note. These figures should be taken as rough approximations only. In Scandinavia in particular herd sizes are often understated.

Appendix 5. Commercial fish catch (including invertebrates), 1953–75 (thousand tonnes)

Country	1953	1962	1965	1970	1971	1972	1973	1974	1975
Northern North Atlantic (ICES sub-areas I, IIb, IIa, Va, XIV, and ICNAF sub-areas 0, 1, 2)									
Norway	915	1,054	1,272	2,074	2,112	2,307	1,982	1,665	1,589
USSR	n.a.	935	846	762	874	698	904	1,236	1,463
Iceland	449	821	1,195	698	634	684	861	901	938
UK	455	457	372	399	331	285	283	287	259
W. Germany	277	387	372	277	259	187	188	264	223
E. Germany	—	—	—	—	—	13	50	125	117
France	57	128	75	69	67	40	34	58	n.a.
Portugal	111	156	134	51	40	28	23	60	34
Denmark (including Greenland and Faeroes)	53	198	147	88	78	73	76	88	107
Poland	—	4	23	75	53	46	21	62	58
Spain	18	60	60	29	42	34	12	34	41
Belgium	20	24	16	15	14	11	8	8	8
Canada	11	—	27	3	5	4	6	3	5
Other member nations	—	—	6	6	2	1	1	—	1
Non-member nations	n.a.	—	64	109	52	—	—	—	—
TOTAL[a]	2,418	4,211	4,611	4,654	4,578	4,411	4,451	4,791	4,845
Northern North Pacific (INPFC Bering Sea and western Aleutian Islands region and northeastern North Pacific sub-areas Shumagin, Chirikof, Kodiak, Yakutat, Southeastern; Okhotsk Sea and adjacent waters)									
Japan	101[b]		508	1,634	2,025	2,144	1,961	1,785	1,409
USSR	343[b]		924	766	1,743	1,615	2,252	2,391	2,588
USA (Alaska)	143[b]		223	242	214	192	194	206	199
Canada	10[b]		20	12	11	9	7	4	3
Republic of Korea	—		—	5	10	13	7	9	7
Poland	—		—	—	—	—	tr.	tr.	4
TOTAL[a]	698		1,675	2,659	4,003	3,973	4,428	4,395	4,210
TOTAL for World Ocean	24,100	40,050	45,570	60,580	59,911	55,132	55,282	58,898	57,989

[a] Details will not always equal total because of rounding. 1953 and 1955 complete catch data not available for Japan and USSR.

[b] 1955.

tr. = trace (less than 500 tonnes).

n.a. = not available.

(*Sources*: USA (Alaska) totals from *Alaska Catch and Production, Commercial Fisheries Statistics*, Juneau, Alaska, Department of Fish and Game; other countries totalled from selected data in Tables 33 and 36, above; USSR totals probably understated.)

Appendix 6. Timber felling, 1960–75 (thousand cu m)

Area	1960	1965	1970	1971	1972	1973	1974	1975
USSR: Far north (defined as 'wood brought out')	28,300	41,000	56,000	58,000	62,000	64,000	66,000	80,000[a]
Canada: Northwest Territories and Yukon Territory (defined as 'wood cut')	162	75	78	71	88	130	140	146
USA: Alaska (defined as 'annual cut')	1,705	2,020	3,060	2,840	2,920	3,160	2,950	2,180
Norway: Nord Norge (defined as 'total round wood cut')	571	456	326	329	354	319	355	390
Sweden: Norrbotten (estimated cut)	2,700	2,800	5,000	4,800	3,489	3,667	3,736	3,586
Finland: Lappi (estimated cut)	7,630	7,090	7,420	7,060	6,890	5,810	6,250	5,040

[a] Area of far north increased by addition of some more territory in Khabarovskiy Kray.

Notes and sources for appendices

USSR

The 'far north' [*krayniy sever*] of the USSR, for which figures are quoted in Appendices 1, 3, and 6, is not exactly defined in the Soviet statistical handbooks, but evidently coincides with the combined 'regions of the far north' and 'localities equated with regions of the far north', as used in the labour legislation (see p. 52 and Map 8). It consists of those administrative areas listed in Appendix 1, *less* Karel'skaya ASSR, and parts of Arkhangel'skaya Oblast' and Komi ASSR, *plus* parts of Tomskaya Oblast', Krasnoyarskiy Kray, Irkutskaya Oblast', Buryatskaya ASSR, Chitinskaya Oblast', Amurskaya Oblast', Khabarovskiy Kray, and Primorskiy Kray, and all of Sakhalinskaya Oblast'. The 'Soviet north' [*sovetskiy sever*], a concept advanced by the economic planner S. V. Slavin in 1958 (see *Problems of the North*, no. 2, 1961, pp. 271–85) includes Karel'skaya ASSR and all of Arkhangel'skaya Oblast' and Komi ASSR, but east of the Urals it has in places a rather more northerly boundary than has the 'far north'. The addition of the more populated western areas gives it a slightly higher density of population. Thus the southern boundaries of the 'far north' and the 'Soviet north' coincide neither with each other nor with major administrative regions for which statistical data are published. The figures quoted for the 'far north' in these appendices are among a small number separately listed in the Soviet statistical handbook *Narodnoye khozyaystvo RSFSR*.

Canada

The 'middle north', 'far north' and 'extreme north' of Canada [*moyen nord, grand nord, extrême nord*] are concepts advanced by L.-E. Hamelin in his *Nordicité Canadienne*, 1974, pp. 107, 136, The 'middle north' includes a broad belt of territory in northern Quebec, Ontario, Manitoba, Saskatchewan, Alberta, and British Columbia.

Principal sources

USSR   *Narodnoye khozyaystvo RSFSR. Statisticheskiy yezhegodnik* (Economy of the RSFSR. Statistical yearbook), Moscow, Tsentral'noye Statisticheskoye Upravleniye.
     Narodnoye khozyaystvo SSSR. Statisticheskiy yezhegodnik (Economy of the USSR. Statistical yearbook), Moscow, Tsentral'noye Statisticheskoye Upravleniye.
     Itogi vsesoyuznoy perepisi naseleniya 1970 goda (Results of the all-Union census of 1970), Moscow, Tsentral'noye Statisticheskoye Upravleniye.

Canada   *Canada Yearbook*, Ottawa, Statistics Canada.
     1976 census. Population: preliminary counts, Ottawa, Statistics Canada.
     North of 60. Oil and gas activities, Ottawa, Department of Indian Affairs and Northern Development. Other statistical material published by this Dept was also used.

Alaska   *Current population estimates by census divisions*, Juneau, Alaska Department of Labor.
     The Alaskan economy. Year-end performance report, Juneau, Alaska Department of Commerce and Economic Development.

Greenland *Grønland. Årsberetning*, Copenhagen, Ministry for Greenland.
Norway   *Statistisk årbok*, Oslo, Central Bureau of Statistics.
Sweden   *Statistisk årsbok för Sverige*, Stockholm, National Central Bureau of Statistics.
Finland   *Statistisk årsbok för Finland*, Helsinki, Central Statistical Office.
Iceland   *Tölfræðihandbók*, Reykjavík, Statistical Bureau of Iceland.
Oceans   *Yearbook of Fishery Statistics*, Rome, Food and Agriculture Organization of the United Nations.
     International Commission for the Northwest Atlantic Fisheries. Statistical Bulletin, Halifax, Nova Scotia.
     Bulletin Statistique des Pêches Maritimes, Charlottenlund, Conseil International pour l'Exploration de la Mer.

Further reading

General

Journals:
Arctic, Montreal, Arctic Institute of North America.
The Beaver, Winnipeg, Hudson's Bay Company.
Grønland, Charlottenlund, Det Grønlandske Selskab.
Inter-Nord, Paris, Centre d'Etudes Arctiques.
Musk-Ox, Saskatoon, Institute of Northern Studies.
Polar Geography, New York, American Geographical Society.
Polar Record, Cambridge, Scott Polar Research Institute.
Soviet Geography. Review and Translation, New York, American Geographical Society.

DARNELL, FRANK (ed.), *Education in the North*, Fairbanks, University of Alaska and Arctic Institute of North America, 1972.
SATER, J. E. (ed.), *The Arctic Basin*, rev. edn, Washington, Arctic Institute of North America, 1969.

Northern USSR

ARMSTRONG, TERENCE, *Russian Settlement in the North*, Cambridge University Press, 1965.
CONOLLY, V., *Beyond the Urals*, London, Oxford University Press, 1967.
CONOLLY, V., *Siberia Today and Tomorrow*, London, Collins, 1975.
DOWLE, D. L. AND KASER, M., *A Methodological Study of the Production of Primary Gold by the Soviet Union*, London, Consolidated Gold Fields, 1974.
D'YAKONOV, F. V. (ed.) AND OTHERS, *Dal'niy vostok. Ekonomiko-geograficheskaya kharakteristika* (The far east. Characteristics of its economic geography), Moscow, Mysl', 1966.

ELLIOTT, IAIN F., *The Soviet Energy Balance*, New York, Praeger, 1974.

GRANIK, G. I. *Ekonomicheskiye problemy razvitiya i razmeshcheniya proizvodi-tel'nykh sil yevropeyskogo severa SSSR* (Economic problems in the development and distribution of productive forces of the European north of the USSR), Moscow, Nauka, 1971.

KIRBY, E. STUART, *The Soviet Far East*, London, Macmillan, 1971.

SHABAD, T., *Basic Industrial Resources of the USSR*, New York, Columbia University Press, 1969.

SHILO, N. A. (ed.), *Sever Dal'nego vostoka* (The north of the far east), Moscow, Nauka, 1970.

SLAVIN, S. V., *The Soviet North. Present Development and Prospects*, Moscow, Progress Publishers, 1972.

Northern Canada

Annual Report of the Commissioner (later *Government*) *of the Northwest Territories*, Ottawa, Yellowknife.

Annual Report of the Commissioner (later *Government*) *of the Yukon Territory*, Ottawa, Whitehorse.

BERGER, THOMAS R., *Northern Frontier Northern Homeland. The Report of the Mackenzie Valley Pipeline Inquiry*, Ottawa, Ministry of Supply and Services, 1977.

BRODY, HUGH, *The People's Land*, Harmondsworth, Penguin, 1975.

JENNESS, DIAMOND, *Eskimo Administration, vol. II Canada*, Montreal, Arctic Institute of North America, 1964 (AINA Technical Paper no. 14).

MACDONALD, R. STJ. (ed.), *The Arctic Frontier*, Toronto University Press, 1966.

REA, K. J., *The Political Economy of the Canadian North*, Toronto University Press, 1968.

REA, K. J., *The Political Economy of Northern Development*, Ottawa, Science Council of Canada, 1976.

ZASLOW, MORRIS, *The Opening of the Canadian North, 1870–1914*, Toronto, MacClelland & Stewart, 1971.

Alaska

ARNOLD, ROBERT, AND OTHERS, *Alaska's Natives and the Land*. Washington, US Government Printing Office, 1968.

GRUENING, ERNEST, *The State of Alaska. A Definitive History of America's Northernmost Frontier*, New York, Random House, 1968.

HARRISON, GORDON S. (ed.), *Alaska Public Policy. Current Problems and Issues*, Fairbanks, Institute of Social, Economic, and Government Research, 1971.

HUNT, WILLIAM R., *Alaska. A Bicentennial History*, New York, Norton, 1976.

ROGERS, GEORGE W. (ed.), *Change in Alaska. Petroleum, People and Politics*, Seattle, University of Washington Press, 1970.

ROGERS, GEORGE W., *The Future of Alaska. Economic Consequences of Statehood*, Baltimore, Johns Hopkins Press, 1962.

SELKREGG, LIDIA L. (ed.), *Alaska Regional Profiles*, Southcentral region, Arctic region, Southwest region, Southeast region, Northwest region, Yukon region, Anchorage, University of Alaska, 1974–7.

WILLIAMS, HOWELL (ed.), *Landscapes of Alaska. Their Geologic Evolution*, Berkeley, University of California Press, 1958.

Greenland

Greenland. Past and Present, Copenhagen, Henriksen, 1970.
Greenland, Copenhagen, Danish Ministry of Foreign Affairs, 1969.
JENNESS, DIAMOND, *Eskimo Administration, vol. IV Greenland*, Montreal, Arctic Institute of North America, 1967 (AINA Technical Paper no. 19).
RINK, HENRY J., *Danish Greenland*, London, King, 1877; reprinted London, Hurst, 1974.

Northeast Atlantic

GREVE, TIM, *Svalbard. Norway in the Arctic Ocean*, Oslo, Grøndahl, 1975.
MCCRIRICK, MARY, *The Icelanders and Their Island*, Denbigh, published by the author, 1976.
MEAD, W. R., *The Scandinavian Northlands*, London, Oxford University Press, 1974.
SØMME, A. (ed.), *A Geography of Norden*, new edn, Oslo, Cappelen, 1968.

The circumpolar oceans

COOLEY, R. A., *Politics and Conservation. The Decline of the Alaska Salmon*, New York, Harper & Row, 1963.
CRUTCHFIELD, J. A. AND PONTECORVO, G., *The Pacific Salmon Fisheries. A study in Irrational Conservation*, Baltimore, Johns Hopkins Press, 1969.
GULLAND, J. A. (ed.), *The Fish Resources of the Ocean*, London, Fishing News Books, 1971.
ROBERTS, B. B., *The Arctic Ocean*, Ditchley Park, 1971 (Ditchley Paper no. 37).
TUSSING, ARLON, MOREHOUSE, THOMAS, AND BABB, JAMES (eds), *Alaska Fisheries Policy. Economics, Resources and Management*, Fairbanks, Institute of Social, Economic, and Government Research, 1972 (ISEGR Report no. 33).

Index

For Product Safety Concerns and Information please contact our
EU representative GPSR@taylorandfrancis.com Taylor & Francis
Verlag GmbH, Kaufingerstraße 24, 80331 München, Germany